THE TIMES

Cryptic Crossword

Book 12

Edited by Richard Browne

Special acknowledgment to David Akenhead for Proof Editing

Published in 2008 by Times Books

HarperCollins*Publishers*
77-85 Fulham Palace Road
London W6 8JB

www.collins.co.uk

9

ISBN 13 978-0-00-723289-5

British Library Cataloguing in Publication Data
A catalogue record for this book is available from the British Library.

Typeset by Susie Bell

Printed and bound in Great Britain by Clays Ltd, St Ives plc

INTRODUCTION

Have you picked this book up on the off-chance, not having done a *Times* crossword before, and wondering if it is for you? If you do word games or crosswords at all, I would encourage you to have a go – the answers are all printed at the back, so you can help yourself along if you get stuck. Indeed, I recommend newcomers not to spend a lot of time being baffled by a clue, but if you find it difficult to get started on a puzzle, just turn up the solution, and read the clues against the answers. You will quickly find that you can understand how quite a lot of the clues work, and probably that a lot of them are nothing like as difficult as they seemed. If some are still baffling, don't worry about them – with more practice their number will diminish, as you start to recognise the cluing tricks and techniques that we use.

As further practical help, I have taken one puzzle and explained it in detail. It uses most of the cluing devices you will encounter in the other puzzles, so with this help I hope newcomers will be encouraged to join in the fun that thousands of addicts already have every day.

This selection of eighty crosswords from *The Times*, all appearing here in book form for the first time, is representative of those that appeared during 2004 and 2005. Unlike the early days, when Adrian Bell compiled most of the puzzles himself, we have a large team of compilers, and the crosswords in this book are by sixteen different people – newcomers (since the last book) John Halpern and Richard Rogan, and existing team members Joyce Cansfield, Dave Crossland, Roy Dean, John Grant, Brian Greer, John Grimshaw, Paul Henderson, Bob Hesketh, Mark Kelmanson, Don Manley, Roger Phillips, Allan Scott, Wadham Sutton, and me. Each of these compilers are represented in this book by at least two puzzles, so there should be a good mix of everything you would expect to find in a *Times* crossword, all done with wit and subtle misdirection, as we try to stop you from seeing too quickly what is actually right in front of your eyes.

Richard Browne
January 2008

A Beginner's Guide to *The Times* Crossword

Across

1 After wrong cut, lad's looking embarrassed (7)
5 Standard of keenness of big name acting in something dirty (7)
9 Frightened, having hinted about subconscious impulses (11)
10 A bit of ballet almost finished (3)
11 Clothing sale — appeal put out for jumble (6)
12 2 — myself and my reflection? (8)
14 Is the map-room's shape changing? (13)
17 The full length of one's notice (9,4)
21 Paper's Platonic leader (8)
23 Maxim accepts English instructions for body-building (6)
25 Journey, missing time and tide (3)
26 Empty-headed husband in tears, being rejected: dirty work! (11)
27 Remove tutu for modelling (4,3)
28 Crucial stage of card game? I'm not impressed (3,4)

Down

1 Show fear of what best man carries in church (6)
2 Dispose of body? I am intervening (7)
3 One very slow European train at first stuck in tunnel (9)
4 Makes mistake, not having children? (4)
5 President needs hand with job (10)
6 Danes routed in this battle (5)
7 Makes use of cider, potentially — I tucked in (7)
8 Scatter brained, is persecuted somewhat (8)
13 With mixed feelings about the environment girl is in (10)
15 Not entirely cheerful conclusion, dropping dead at last, in the event (9)
16 In poor accommodation under a month, not one right for her (8)
18 One seeking fur turned up material piece (7)
19 Baby produces gas, having swallowed (7)
20 Time to relax on top of lounger — it helps us to nap (6)
22 Music is constant during party (5)
24 Say, stick in the post (4)

NOTES

This detailed explanation of the cluing methods for this puzzle should show you plenty of techniques (for example, some abbreviations) that are used in most of the other crosswords. I have indicated the definition elements in bold type: you will notice that virtually every clue has both a definition and a cryptic element (so these two routes to the answer should enable you to crosscheck your answer and be confident it is correct); and that the definition always comes either at the start or the end of the clue (useful tip!).

Across

1 After wrong cut (CRIME, minus last letter), lad's (SON is) **looking embarrassed** (7) = CRIMSON (*is* means *this wordplay leads to the answer*)

5 **Standard of keenness** (reference to *keen as mustard*) of big name acting (STAR) in (inside) something dirty (MUD) (7) = MUSTARD (*of* means *is made up from*). Notice the pun on *acting*

9 **Frightened**, having hinted (INTIMATED) about subconscious impulses (ID – Freudian psychology) (11) = INTIMIDATED

10 **A bit of ballet** (PAS, a ballet step) almost finished (PAST, minus last letter) (3) = PAS

11 Clothing (GARB) sale — appeal put out (SA, or *sex appeal* – a common crossword conceit – is removed from SALE) for **jumble** (6) = GARBLE

12 **2** (the answer to clue 2 (down) INTERIM is the definition) — myself and my reflection? (ME and ANTI-ME!) (8) = MEANTIME

14 Is the map-room's **shape changing**? (anagram of *is the map-room's*) (13) = METAMORPHOSIS. Here, the definition and the anagram indicator are one and the same

17 **The full length of one's notice** (9,4) = ATTENTION SPAN: the whole clue is a deceptive definition, nothing to do with leaving a job!

21 **Paper's Platonic leader** (8) = GUARDIAN. Two definitions, one a reference to the Guardians who Plato recommended should be the leaders of his ideal state (if one doesn't know this classical reference, the clue is still solvable from the other definition)

23 Maxim (GNOME, a precept or pithy saying) accepts English (*English* is commonly used as code for the letter E, for which it can stand) **instructions for body-building** (6) = GENOME (the E goes inside GNOME)

25 Journey, missing time (TRIP, missing T, another common abbreviation) and **tide** (3) = RIP

26 Empty-headed (INANE) husband (H, another abbreviation) in tears (inside SNAGS), being rejected (all turned round): **dirty work!** (11) = SHENANIGANS

27 **Remove** tutu for modelling (anagram of *tutu for*) (4,3) = TURF OUT

28 **Crucial stage of card game? I'm not impressed** (3,4) = BIG DEAL (two definitions)

Down

1 **Show fear** of what best man carries (RING) in church (CE, another abbreviation: CH is another shorthand for *church*) (6) = CRINGE

2 Dispose of body? (INTER) I am (I'M) **intervening** (7) = INTERIM

3 **One very slow** European (E) train at first (T: this is a common way of indicating the initial letter of a word) stuck in tunnel (inside SIMPLON) (9) = SIMPLETON. A typical *Times* clue, in that the definition is not at all obvious from the wording of the clue

4 **Makes mistake**, not having children? (NO D, S – abbreviations for *daughter, son*) (4) = NODS

5 **President** needs hand (MITT) with job (ERRAND) (10) = François MITTERRAND, late President of France

6 Danes routed in **this battle** (5) = SEDAN, scene of battles in 1870, 1940; anagram of DANES

7 **Makes use of** cider, potentially (APPLES) — I tucked in (the letter "I" put inside) (7) = APPLIES

8 **Scatter** brained, is persecuted somewhat (some consecutive letters from "braine*d is perse*cuted") (8) = DISPERSE. (It can be difficult to see that *scatter brained* is not to be read together; the compiler has omitted the customary hyphen, which would be unfair here)

13 **With mixed feelings** about the environment (AMBIENT) girl is in (VAL inside) (10) = AMBIVALENT

15 Not entirely cheerful (HAPPY, without its final Y) conclusion (ENDING), dropping dead at last (taking off the last letter of *dead*, D), in **the event** (9) = HAPPENING

16 In poor accommodation (GARRET) under a month (MAR, abbreviation for *March*), not one right (R stands for *right*, and we must find one of the Rs to remove) for **her** (8) = MARGARET

18 **One seeking fur** turned up (reversed the letters of) material (REP, a ribbed fabric) piece (PART) (7) = TRAPPER

19 **Baby** produces gas (NEON), having swallowed (ATE) (7) = NEONATE

20 Time (T) to relax (EASE) on top of lounger (L) — **it helps us to nap** (6) = TEASEL (a prickly flower-head used to raise a nap)

22 **Music** is (IS) constant (C) during party (inside DO) (5) = DISCO

24 Say, stick (JAM, which sounds the same) in **the post** (4) = JAMB

THE PUZZLES

1

1

Across

1 Ill-bred type once inspiring love song (5)
4 Vague summary this compiler's beginning to enthuse about (9)
9 Smuggled liquor? Nonsense (9)
10 Empty pub, seizing a hallucinatory drug (5)
11 Church in Glasgow park lacking right kitchen appliance? (6)
12 Classification of duty levied in respect of old yacht (8)
14 Just claim by Windsor gentleman at start of book? (5,4)
16 She inspired poetry a long time before rejecting Holy Writ (5)
17 Alternative notes a Gaelic bard abridged (5)
19 Organised action in a plant (9)
21 Listener catches girl with unknown diplomat (8)
22 Healthy fellow provided with suitable equipment (6)
25 Group of stars spouting tediously about port (5)
26 Playful like a queen's offspring? (9)
27 PM meets resistance in city —it may be a bitter blow (9)
28 English head going over German industrial centre (5)

Down

1 Enter single broadcast and achieve recognition (4,4,4,3)
2 River horse, say? (5)
3 Supple plant raised in fiction (7)
4 Islands to south affording shelter for black bird (4)
5 Superiority gives power in relation to hanging (10)
6 Summary of Erasmus's first religious book (7)
7 Lover a Catholic picked up in Italian area (9)
8 We envy her, and not surprisingly, at times (5,3,3,4)
13 Man cleaned out back of car for uncle, once (10)
15 Report of peevish judge upset about current witness (9)
18 Shrubby plant essential to Arabs, in theory (7)
20 Warship? It capsized taking soldiers across river (7)
23 Works hard making net (5)
24 Defer time inside, for example (4)

1

2

Across

1 Hot air temperature accompanies graceless walk (7)
5 Tropical food idiot dipped in wine (7)
9 Criminal who knows way round court, you might say? (9)
10 More to score in order to obtain drug (5)
11 Hysterical, given face very painful (4-9)
13 Determined man will briefly be needing good books (8)
15 Part of engine —from town in Wales, get one for free (3,3)
17 Document result of firing? (6)
19 Capital against cutting pudding (8)
22 Competition entered, hay and vet ordered? (5-3,5)
25 Here, refuse to leave house (5)
26 In fish, means of identifying criminal, such as him? (9)
27 Clear up, having undergone cosmetic surgery? (7)
28 Amateur admitted to grass being passed on (7)

Down

1 Wrong run going uphill (4)
2 Charge account with copper, shilling and a pound (7)
3 Fear still grips the right (5)
4 Novelist losing way offered grand service (8)
5 The Magic Roundabout? (6)
6 Final candidates offered strong drink keel over (9)
7 Like single figure, simple (7)
8 Prepare old Quixote, perhaps, for battle (10)
12 Thoughtful message carrying faithful instruction in bonnet (10)
14 A Catholic involved in rebuilding a noble city (9)
16 Shrub of which a small number planted in bank (8)
18 Vegetable cut, no point having to cut again (7)
20 Orphan from Stowe holding up a university's final examination (7)
21 Cake provided, but not quite finding space (6)
23 Message weak, not written down, one received (1-4)
24 Good, free crossword, but no clues (4)

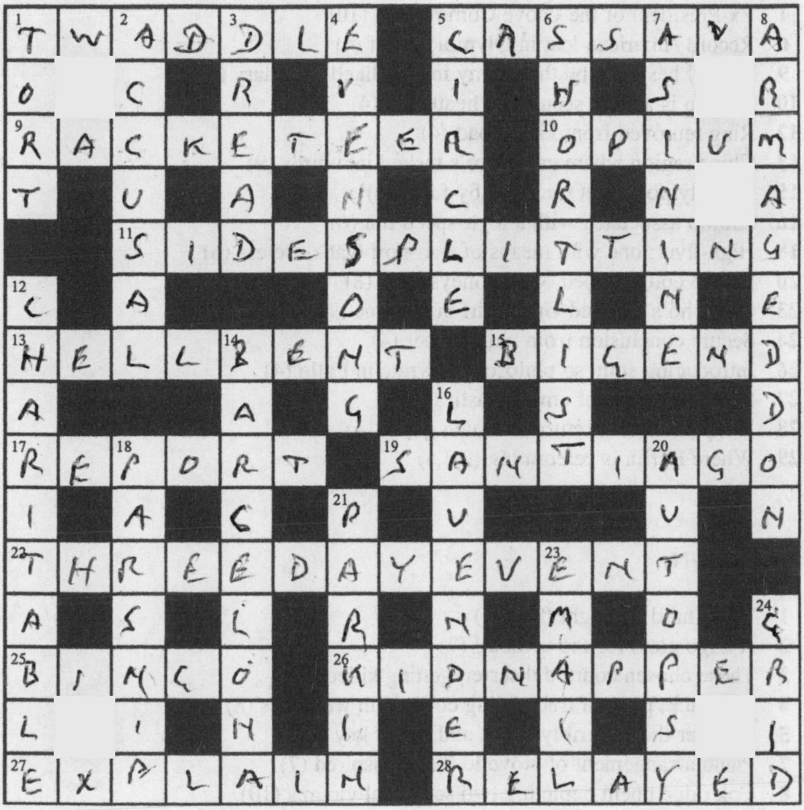

2

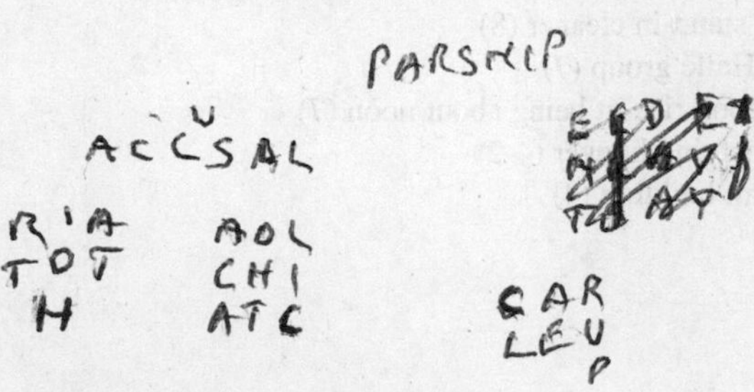

3

Across

1 Ex-President of the Glove Commission (10)
6 Record American lost in Olympic event (4)
9 Capital besieged by the enemy in woodland boundary (6,4)
10 Othello is also to secure the heathland (4)
12 Ring removed from kitty's pad (4)
13 China region where carnivore's tucked into birds (9)
15 Awfully poor duet removed by force (8)
16 Hilltop associated with a high-speed run (6)
18 High-flyer: one with means of transport that's useless (6)
20 Beside court, in bed, wild honeysuckle (8)
23 One who's adopted Greek girl in his time (9)
24 Secure conclusion from philosopher (4)
26 Introducing sum, so philosopher wrote in Latin (4)
27 Source of laughter among rustics? (10)
28 Weapon ensures return of stolen goods (4)
29 Where Berlin is celebrating (2,3,5)

Down

1 Test's hard for night flyer (4)
2 A large glass for the acrobat! (7)
3 Those chosen courted disaster, getting killed (12)
4 Fell back, parts of track being covered in wet grass (8)
5 Number that's terribly tinny, and out of key (6)
7 Piano arrangement of Novello that is inspired (7)
8 Coleridge poem depicting two scriptural victims (10)
11 Well-earned money? (12)
14 This might consist of nine thumps (10)
17 Pay city something for stand-in cleaner (8)
19 Brisk movement from Hallé group (7)
21 He encourages one to cook rice, it being about noon (7)
22 Substitute not quite satisfying hunger (4-2)
25 Purchaser heard for farm building (4)

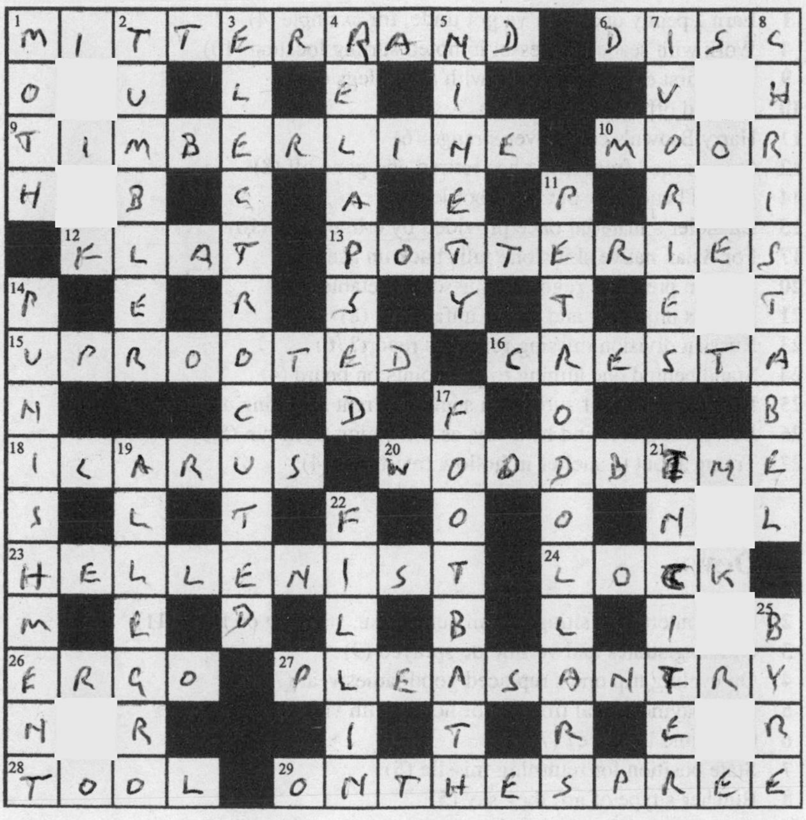

4

Across

1 Earn a penny once you've got trade, for example (4)
4 Work with seabound vessel in noted surfing location (10)
9 Blow first of kisses to miss with curvy legs (5-5)
10 Cut end off shaving stick (4)
11 Harry Brown's come over strange (6)
12 One rescued from firing has retired and gone off (8)
14 Armed band clear out from borders (4)
15 Chandler's material once provided by crime spate (10)
17 For Asian native skills one falls back on ace (10)
20 Back in pre-teens regularly missed vegetable (4)
21 Phrases unjoined, and as yet unfamiliar (8)
23 Russian division missing referee's report? (6)
24 Stand behind one aiming to put points on board (4)
25 I abandon heart of subject in semi-coherent summing up maybe (10)
26 Belt tucked in round hen, say, as protection from sun (5,5)
27 Tramp about to shelter in hollow tree-trunk (4)

Down

2 Fools entertain visiting Italian gentleman, standing on head (11)
3 Old magistrates had vermicide sprayed (9)
4 On cycling trip one's replaced good ladieswear (7)
5 If no payment, real trouble for house with vacancy (3-6,6)
6 Core time in office? (7)
7 State position for returning missile (5)
8 Butcher's type of art, they say (5)
13 One making hollow bargain over drug with urchin (11)
16 Joining study on illness at hospital department (9)
18 In one visit, keys drop in glacier (7)
19 A different girl? (7)
21 First of autumn leaves in stacks (1,4)
22 Brute's expression of delight (5)

4

5

Across

1 Leave out bar (6)
4 Supporter relaxed about quiet senior officer (5,3)
10 Bury money, say, produced within thc family (9)
11 Author in study to finish off tale (5)
12 Removal of cycles by river (7)
13 Flower's English name is wrong (7)
14 Tasty food that goes hard (5)
15 Sergeant-major with mobile blowing up obstructive clot (8)
18 Mother's almost divine butcher (8)
20 Binding a hat with velvet initially (5)
23 Cause damage, roughly pulling tail off stork (7)
25 Farewell to a Liberal set (7)
26 Most prominent star's expression of triumph about record (5)
27 Ditch insect is biting (9)
28 One produces notes, which is very important to the directors (8)
29 Webster noticed, it's said, by monarch (6)

Down

1 Data for study in English immorality (8)
2 Straight and narrow path for model girls (7)
3 Taps with the finger in company, upset by bad language (9)
5 Feature of a male habit that sets brides cheering madly (6,8)
6 Milling crowd that's on the booze (5)
7 Judge's directions for funeral bearers (7)
8 Enclosure for little piggies (6)
9 Film conflict, getting lawyer over (5,9)
16 Friendly vessel, well-handled, one passes (6,3)
17 Admirer I subsequently cheat (8)
19 Degeneration of a scalp (7)
21 General hospital room in the Islands (7)
22 Board's rating covers business degree (6)
24 Vote against party (5)

5

6

ACROSS

1 Creature a soldier captures abroad (6)
5 Bring to a conclusion nevertheless, proceeding very quickly (6,2)
9 One authorises girlfriend briefly to enter house (8)
10 A little bit close to northern resort (6)
11 Academic qualification obtained from inside Greece (6)
12 Paints with colour — that'll give an edge (8)
14 Labour schedules about a day to go in pursuit of Tory voters (12)
17 Dress — terribly swish attire, right? (12)
20 Henry and William given notice (8)
22 Planning itinerary right away for excursion (6)
23 Broken lances in old battle site (6)
25 Class makes gradual progress, and finally settles (8)
26 Rats consuming dead fish (8)
27 Gift of story book (6)

DOWN

2 Red or white wine, good at parties (6)
3 Stressed no more than nineteen qualified at last (11)
4 Aware Keith somehow has inscribed name on top of the present (2,3,4)
5 Spell inhaling pot leads to physical collapse (7)
6 Cloth to flog after cutting (5)
7 Orthodox leader, one limiting black magic (3)
8 City dweller thrown out by a tribune (8)
13 I was put in with the rest into the exam, not the practical (11)
15 Designer also introduced by film director on set out east (9)
16 *Stormy Weather* has a hold on a singer (8)
18 Trouble one large town almost completely banned (7)
19 Understood, getting to the end of a French translation (6)
21 Preserved, like former Peruvian empire (5)
24 Gigot is part of course (3)

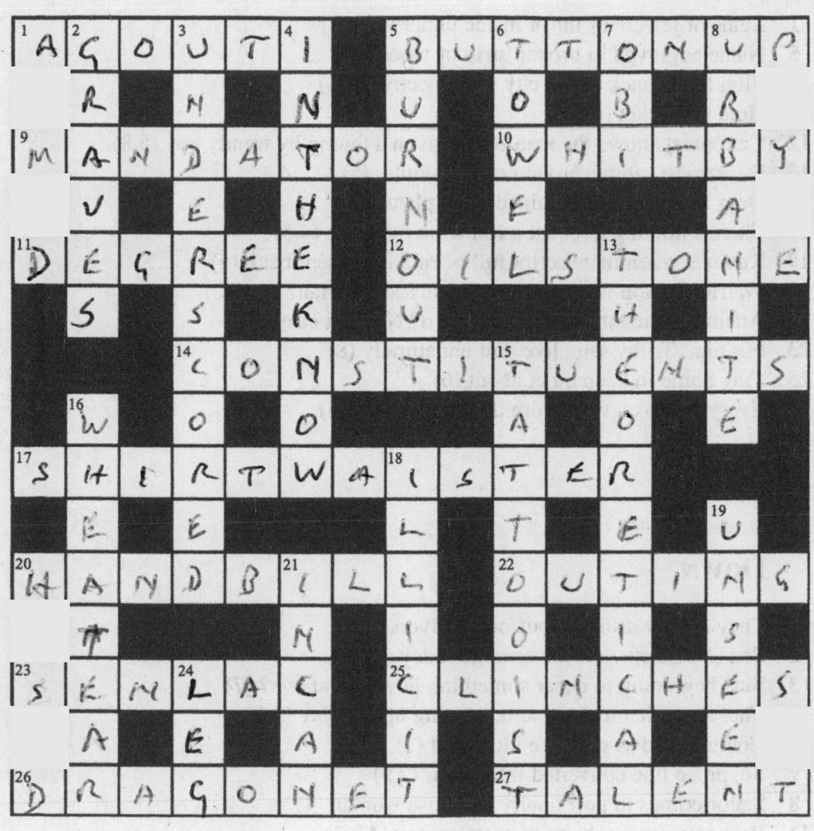

7

ACROSS

1 Items of jewellery finish inside underwear (8)
5 Name engraved in broken strip of wood (6)
9 Film looks back to old city retrospectively (8)
10 Hold up basket (6)
12 Cartoonist shows the man being given a thump by trendy boy (5,8)
15 Deck sport returns around end of winter (5)
16 Drug satisfactorily obtained from plant (9)
17 There's nothing to equal a romantic marriage (4,5)
19 Rumpus when manuscript full of rubbish is sent back (5)
20 Warrior bishop who wrote poetically about Man? (9,4)
22 Attempt to arrest bumpkin not hard? Not that easy (6)
23 For him, Trilby sang live, but unnaturally (8)
25 Tiny home to keep quiet about (6)
26 Everyone has a wish from the beginning (3,5)

DOWN

1 They supply drink — buffoon delivers (10)
2 Son should stay off the cocaine today (3)
3 Side beginning to cheer something hovering above? (7)
4 Those people on river, some turning up for play (3,9)
6 Commended as pressure increased (7)
7 So, prose line converted into poem (2,9)
8 Sailor comes to port finally for loose woman (4)
11 Be a satanist maybe with great energy (4,3,5)
13 At one villa I failed to offer relief (11)
14 Tramp in grass asleep (10)
18 Wild animal? Only Pussy, we hear (7)
19 Is angry, possibly becoming lilac? (7)
21 Saint, to stomach unfermented grape juice (4)
24 Very eager not to finish earlier (3)

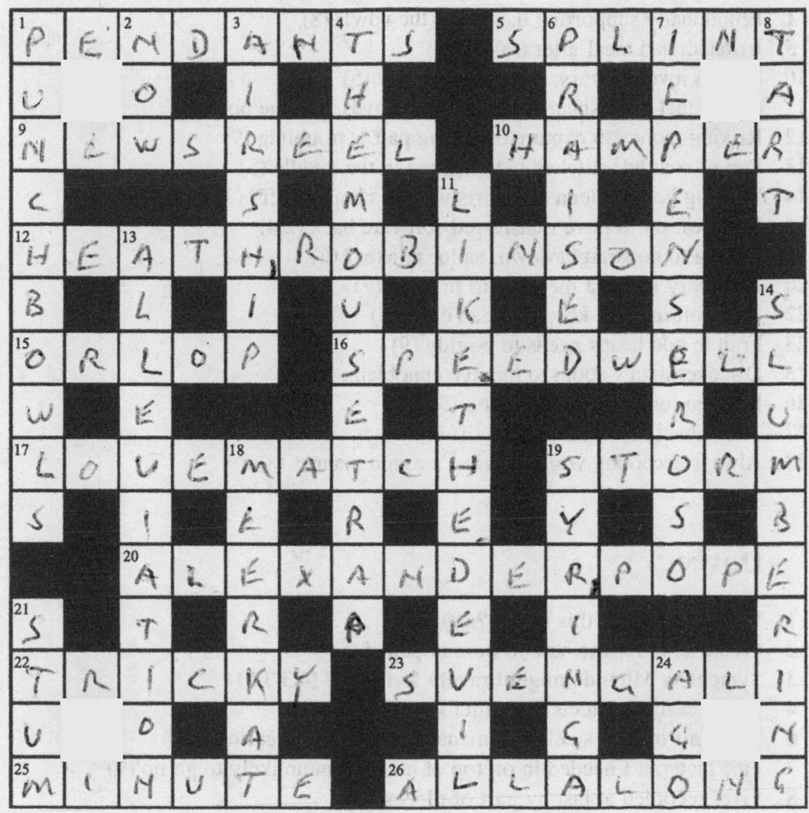

8

ACROSS

1 Emotionally supporting daughter (the cow!) (8)
5 Irritated, had meal after tea (6)
10 Fruit is revolutionary, but rotten inside? (5)
11 Son to feel enthusiasm dipping into Graves, say, the poet (9)
12 Risking money to capture opposing pair is troubling (9)
13 Part of coat has stitched tag changed in the middle (5)
14 Floating beam Queen of Hearts wanted chopped off? (7)
16 Unnecessary to have transferred software back (2,4)
18 Having to go straight down, sailor relaxes? (6)
20 It's Henry I send a message to in town (7)
22 Let's roll in mud, keeping head clear (5)
23 Fruit in odd heaps needs to be tidy (9)
25 Discover man's about to receive ennoblement? (9)
26 In Open round, reaching green (5)
27 To retain staff, earnings take off (6)
28 Always accepting worried priest's rough manner (8)

DOWN

1 Whale seen from this bridge? (8)
2 Characters in name sure to be misspelled (5)
3 Symphony Miranda might bravely welcome? (4,3,3,5)
4 Tough casing protects computer key, *Cancel* (7)
6 With half of pipe steel cast, manage far-seeing invention (6,9)
7 Fine material I needed in on top of house: I'm unlikely to go up (9)
8 Lives recorded at last by part of diocese (6)
9 Not fatally wounded? So like Cupid (6)
15 Ordering equipment banned after speaking (9)
17 Like a master, so no longer showing insolence (8)
19 Dog Star (6)
20 Anger suitors extremely as he takes the lead for Portia, say (7)
21 Poor foreshore road under repair (4-2)
24 Countryman needing map covering Ireland (not the north) (5)

9

Across

1 Lie about origin of recurrent animosity (8)
5 Without hesitation, Australian friend catches partners in trap (6)
9 Backing given to blood relative for a second time (3)
10 She was spotted sleepwalking (4,7)
12 Obscure knight accepted in fashionable area right away (10)
13 The first person who transformed the Palladian style? (4)
15 Dagger rarely found on cricket pitch (6)
16 Serious attention given to machine-gun emplacement (7)
18 Pupils' examiner has place in old sect (7)
20 Standard quarters adjoining house (6)
23 Muslim ruler's poem once rejected (4)
24 Doomed American state played a leading role (3-7)
26 Cutter has power to move quickly round lakes (11)
27 Drink, it's said, produces trouble (3)
28 Part of musical comparatively lacking in feeling (6)
29 Ruffian in film deceived about drug (8)

Down

1 Marvellous husband leaves sumptuous material (6)
2 Member of popular team getting runs (7)
3 Clan's aim? It worked like a charm (10)
4 Implausible relations of senior union members? (3,5,5)
6 The last word in fairy stories? Quite the opposite (4)
7 With revolutionary supporters, duke goes up to cajole (7)
8 A monster, even though he's a colleague of 5 (8)
11 Knife used before singer consumes extremely smooth fish? (8,5)
14 Perhaps crab tin contains sign of decay: something one can't return (10)
17 Unsophisticated old Foreign Secretary's play on words (8)
19 Regular group of students on degree course? (7)
21 Refuse to keep key in estate accommodation (7)
22 Welshman holds theologian to be mixed-up (6)
25 Turn up with soldiers to get monster (4)

10

Across

1 Manor erected by old money, guarantees of quality (9)
6 Seedy operative may get worse (5)
9 Return from bar to party (5)
10 Like chimps? Chap can catch a million, going to African country (9)
11 Article put in containers by army: this needs two packs (7)
12 Touching, Samuel's display of conscience (7)
13 Fundamental course that may get you set up for life (10,4)
17 Its doors are open daily in the lead-up to Christmas (6,8)
21 Spell with endless suffering: there are several sides to it (7)
23 Overdue book was stimulating (7)
25 The worst nettles may be very near (9)
26 Definitely confirm one's at home with family (3,2)
27 Countryman's last to call on union (5)
28 Where the only way is down, as Pooh found? (5,4)

Down

1 Author about to receive printout (4,4)
2 Spice up vile new concoction (5)
3 Full of woes, see me not shifting responsibility (9)
4 French poet sounds a mindless brute (7)
5 Biblical land has appeal to girl I've just met (7)
6 Mass fight contained by NCO (5)
7 Cat has three minutes with boxer collection (4-5)
8 Smelling worse in private (6)
14 Try to preserve value of finger ring (5-4)
15 Sounding a little tipsy, disregarding any other seafood (9)
16 Caution: one easily shocked about enclosure (8)
18 Time to lock up old abbey site (7)
19 He decides, having arrived, to eat snack (7)
20 Why king is drunk? (6)
22 Novelist making huge amount? Knock one zero off! (5)
24 City in Kentucky, also outside (5)

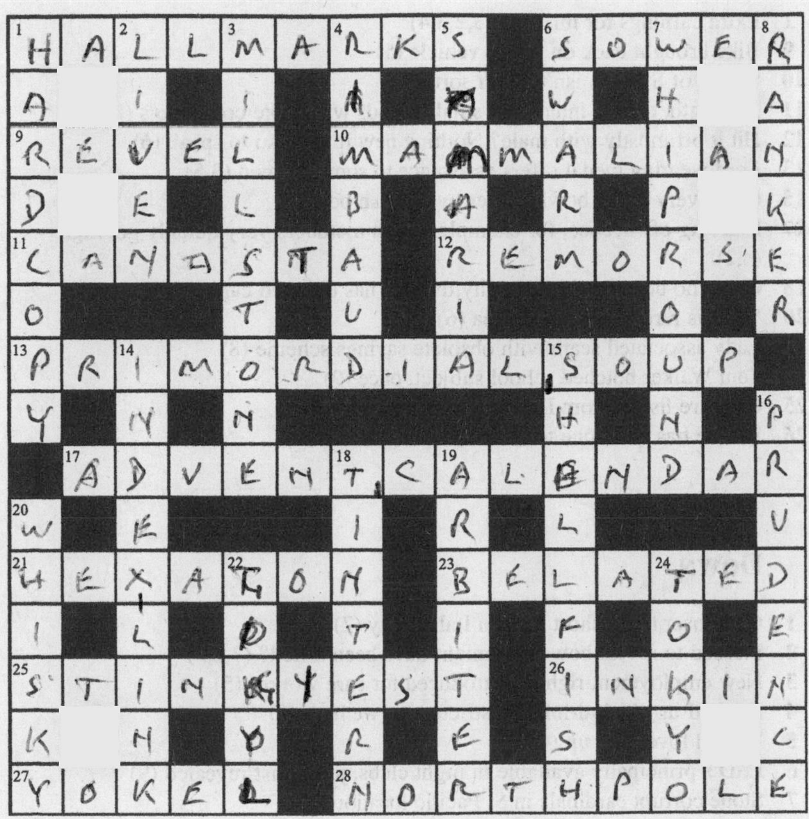

11

Across

1 Extra earnings for mistress (3,2,3,4)
9 Bike brought back on motor vehicle (5)
10 Queen of Sweden isn't a rich sort (9)
11 At length, people inhabiting small islands will make complaints (8)
12 Hit it off mostly with male? Nothing new in that, so to speak (6)
13 Lord wearing medal offers assistance to some women (3,5)
15 One's very much bowled over by an Irish poet (6)
17 Entering off-licence, for example, union members very quickly got high (4,2)
18 One who doesn't object as city dweller has doubled capital? (8)
20 Wind is rare in mountain area (6)
21 Lady associated scam with obsolete savings scheme (8)
24 Tom Walker botched school subject, once (9)
25 Creature fished from Danube, astonishingly (5)
26 He/she has good one to finish drinking bout (6,6)

Down

1 Composer has a short time in Italian city (7)
2 Pleased to reveal how trout might have been killed? (7,2,5)
3 New employment right is introduced for care worker (5)
4 Festival in which drink is restricted, so we hear (8)
5 Wound lover boy up (4)
6 LSD's principally available in night clubs, journalist revealed (9)
7 Stone corrupt cardinals in S. Pacific location (8,6)
8 After short dramatic scene, thespian finally entered the stage (4,2)
14 Answer question artist raised over French woman's painting technique (9)
16 Fellow labourer having pig initially replaced by bovine creature (2-6)
17 Spy a fellow's brought in as a plant (6)
19 Oven-ready bird daughter removed from car (7)
22 Story book kept inside an item of furniture (5)
23 Obedient hound, perhaps, trained up to catch duck (4)

11

12

Across

1 Big exhibition attracting inevitable publicity (8)
5 Another messenger? That covers him (6)
9 Shares wine with pages (9)
11 Last decoration on roof I removed (5)
12 You have to steer deviously to find favour (7)
13 Liking to appear penniless arriving at entrance (7)
14 Hard copy about an orchid translated by Molière's eponym (13)
16 Showing little enthusiasm, so 19 would be beaten (4-9)
20 Tug from child grabbing women's scarf (7)
21 Time to get hold of aural pornography (7)
23 Former partner's left — about to celebrate (5)
24 Once African island visited by Scot (9)
25 Rugby players quietly indulge in theft (6)
26 Mentioned Bill and Edward (8)

Down

1 English politician with a passion for power (6)
2 Current value put on one's capital (5)
3 Car, second time after time, finally first (4-3)
4 Exciting ride in vessel overwhelmed by large wave (6,7)
6 Expunged a handful of notes (7)
7 Lying people start to deceive a whole town (9)
8 Costliest production losing a little time: *The Longest Day* (8)
10 Last swallow departing? (3,3,3,4)
14 Jug is meant to stand up (4,5)
15 Let's chat about personal property (8)
17 Nasty mould discovered under house — tough! (7)
18 See 22's heart stop — heart is removed (7)
19 A lot of beer is drunk (6)
22 Ignoramus's ignorance doesn't irritate ordinary teachers, only Heads (5)

12

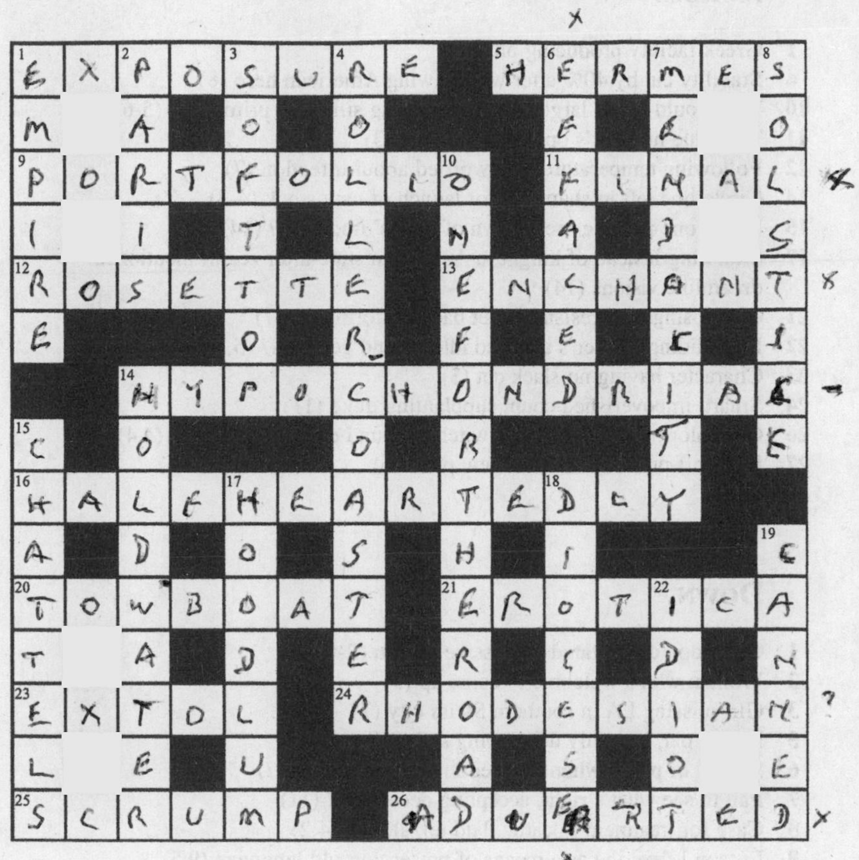

13

Across

1 Greek facility producing oil (6)
4 Stability cut by 40% after withdrawing American hero (8)
10 Who could shape large bowls containing silicates, primarily? (5-6)
11 This clue number's opposite number? (3)
12 Following temperature, star worried about infection (7)
14 Circle line left in shambles for launch of new stock (4-3)
15 Might one provide breakdown of oldies' impetigo? (14)
17 Shot king instead of knight in Australian state after zealot mentioned dreamlike visions (14)
21 Grid losing core resistance set back performance (7)
22 Backsliding boozer's about to idle — and get fired? (5,2)
23 Character having no slack cut (3)
24 Smart, impoverished count supplanting duke (11)
26 Old colony making central switch to sound of horn and gong (4,4)
27 Head off provocation over new plot? (6)

Down

1 Giant dog decapitated aggressive person (2-6)
2 Woman who's welcome to come up (3)
3 Girl missing LA in southern Swiss city (7)
5 Below par, I'm duly developing so? (8,6)
6 Source of peace when inserted in hearing organs (7)
7 Fail to see what's right, accepting deception? (11)
8 Case for treatment in cattle, laid up, almost (4,2)
9 Lesson I describe as a means of preserving old language (9,5)
13 A German I twice introduced to dotty aunt, making meditative movements (3,3,5)
16 Tsar up in rages because of him? (8)
18 Criticise arrest involving tip off (7)
19 Allow area round island's rough ice (7)
20 With a change of hands, grab fork (6)
25 Result of nurse ignoring onset? (3)

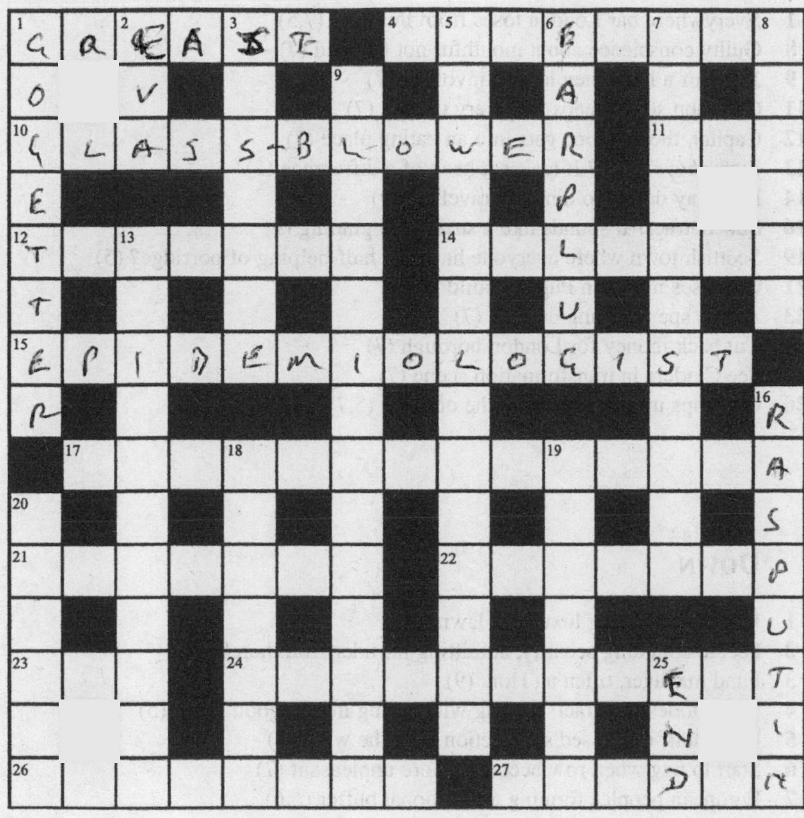

14

Across

1 Everywhere bar London loses from this tax? (7,5)
8 Guilty conscience about mouthful not finished (7)
9 Jewel of a lass when love is involved (7)
11 Common sense keeps king very uneasy (7)
12 Capital, the way one gets into an eating place (7)
13 Sign *above* letter? It makes a heck of a difference (5)
14 Highway deeds holding up travellers (9)
16 Subcontract? It sounds like a striking beginning (9)
19 Scottish town where everyone has only half-helping of porridge? (5)
21 Criticises marks in Pugin's building (7)
23 A time spent getting dressed (7)
24 Cut back money for London borough (7)
25 See Cinders in transformation scene (7)
26 Girl pops up somewhere in the outback (5,7)

Down

1 University cutting luxurious lawn (7)
2 Feeble-sounding security, admitting intruder from here? (7)
3 Fund manager, often an Hon. (9)
4 Very modest contract: leading with a king makes good sense (5)
5 Plane staff expressed satisfaction with the wind (7)
6 Start to nag when row becomes more unpleasant (7)
7 European peoples forming a traditional buffer (6,6)
10 Abandoned candidature, being too far behind (4,8)
15 The boy with 'er can't compare with the British Grenadier (9)
17 Disorder in the Capitol is newsworthy (7)
18 Fundamental sort of farming (7)
19 Worker is a new supporter of culture (7)
20 Big ship, the *Bounty* (7)
22 They make it difficult to see in a dusty escritoire (5)

14

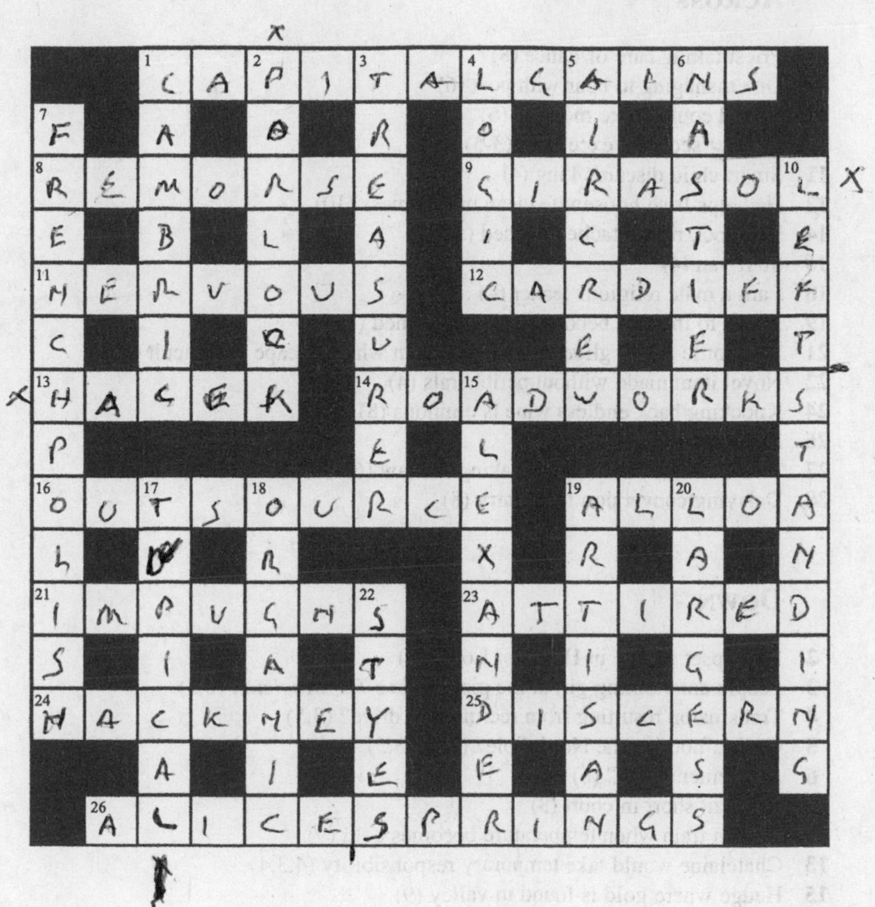

15

Across

1 Priest taking care of dance (8)
6 One managing to bear with poet (6)
9 Mould could make me wild (6)
10 Highest score — excellent! (3-5)
11 Smart child discards king (4)
12 He helps Irish housewife clean up her mess (10)
14 Deadlock man attached to shed (5-3)
16 Start van (4)
18 I am a male religious leader (4)
19 Listen to the row becoming less restrained (8)
21 Bluebottle has to glide into a place from which escape is difficult (10)
22 Novel item made without peripherals (4)
24 Knocking back endless wine is ominous (8)
26 Plot crushed (6)
27 This can be achieved by breaking the law (6)
28 Delaying conversion to idolatry (8)

Down

2 Port upset couple in House of Lords (5)
3 People entertaining girl at the pictures see *The Magician* (8,3)
4 Transfusion resulting from recruitment drive? (3,5)
5 Cock-a-hoop at the North Pole? (2,3,2,3,5)
6 An element of PC (6)
7 Yarn cut short in court (3)
8 Egg on train when temperature becomes cold (9)
13 Chatelaine would take temporary responsibility (4,3,4)
15 Hedge where gold is found in valley (9)
17 Mother has to dress up little boy showing glee (8)
20 Malfunction of arc light, flickering when deprived of argon (6)
23 Secondary road heading north followed by soldiers (5)
25 Bad state (3)

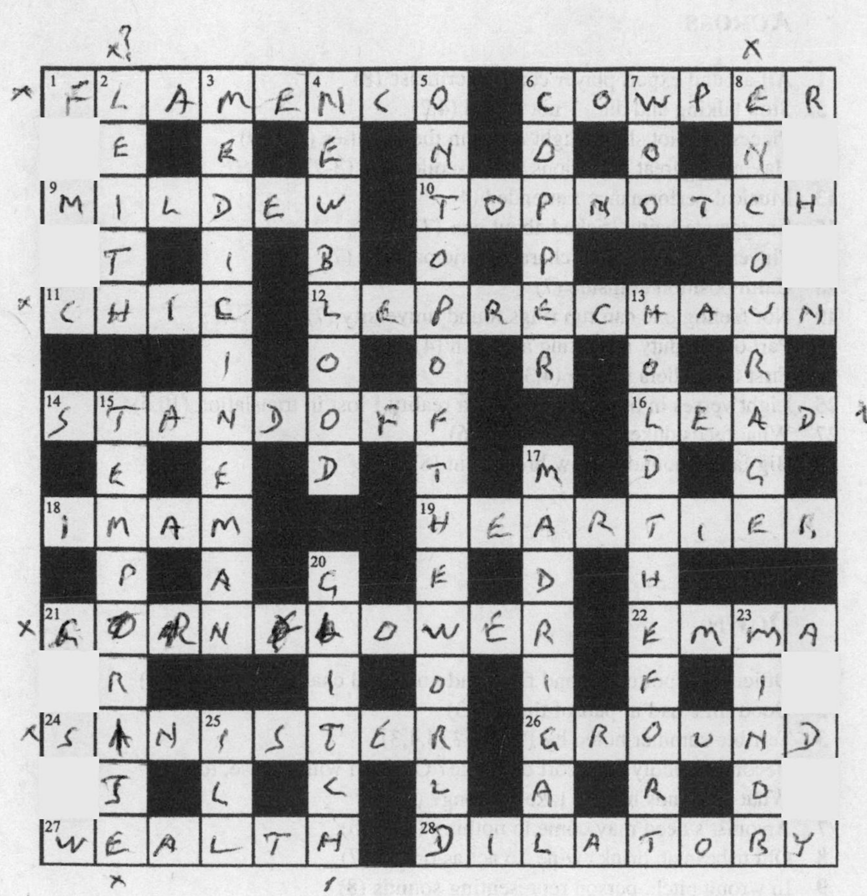

16

Across

1 All around expert player come to criticise (8)
5 Stop talking and die — not in bed (4,2)
10 Pieces of Eliot she thought novel in the twenties (2,3,10)
11 Having no great objections to egg-collecting (3-7)
13 Musical performance I attended (4)
15 Exaggerate being tricked about gun (7)
17 Players cooperating in characteristic passage (7)
18 Calm position at inside (7)
19 Not feeling one can run rings round university (7)
21 Part of our duty is making a speech (4)
22 First two fallers in race (4,3,3)
25 Light verses in books concealing a reality I lost in translation (10,5)
27 What feeds duke, and this man? (6)
28 Big fall in pound — row breaks out (8)

Down

1 Officially reported second river had gone and changed channels (7)
2 Good time had in part of theatre (3)
3 Terrible summer noted by Puccini? (3,4,3)
4 Become friendly with sort of beetle? One can with mouse, too (5)
6 What one finds hard to take for long? (4)
7 Arsonist's deed may come to nothing (2,2,2,5)
8 Old tribesman drinks wine, to act as oracle (7)
9 In wrong pitch, person representing sounds (8)
12 After drunk turns up on beach, go off to rest uneasily (4,3,4)
14 Loved one's not up in town (10)
16 Not looking forward to daughter going to university? (8)
18 Moved fast to distribute money round church (7)
20 Poet's sister absorbs as much Pascal as possible (7)
23 Refuse old gondolier (5)
24 Expression of disgust about head of foreign kingdom (4)
26 Appear to lose king in card game (3)

16

17

Across

1 Force a small child to speak (5)
4 Non-stop walkers (7-2)
9 Player on the side getting ready? (9)
10 Form of communication between male and female, we hear (1-4)
11 Athlete light in comparison with opposition (9,4)
14 Survive after all the others (4)
15 Picked from thousands of orthodontists, etc. (3,2,5)
18 Rough value of company I partner with (10)
19 Consequently turned into monster (4)
21 Beginning of sentence landlord got after serious kind of offence (7,6)
24 In favour of restricting publicity for museum (5)
25 Special pride's associated with computer network that can catch bugs (6,3)
27 Release revolutionary design ahead of time (9)
28 With central India very warm, one is put on (5)

Down

1 Scholarly analysis involving scores of masters (10)
2 Attempt a stately annexation of India (3)
3 Drinks dispenser of unusual nature (3-3)
4 Change used on my novel? (9)
5 Material one's taken from Rachmaninov, ultimately (5)
6 Order that stops men from taking the right view (4,4)
7 Coach runs into conflict producing problem for players (5,6)
8 Initially, you expect little puppy to bark sharply (4)
12 Society's top people providing high-level instruction (11)
13 Church that is supported by monk, bishops, and current religious leader (5,5)
16 Settled learner amongst top people as subservient follower (9)
17 Get off one's soap-box and resign (4,4)
20 Suffer detention, watched (6)
22 Match a protective coat (5)
23 A page editor cut was imitative (4)
26 How to get wife with rings? (3)

17

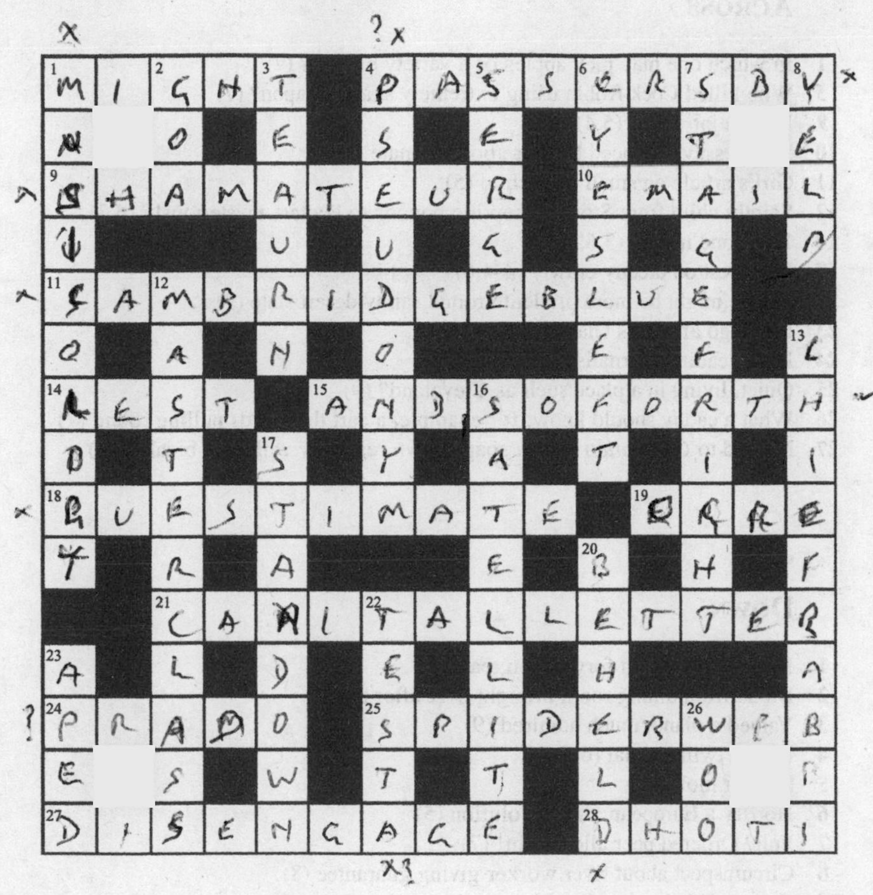

18

Across

1 In which one may pick apples or a variety of beet? (7)
5 Who killed Cock Robin using extremely sharp weapon? (7)
9 Mark's job in tax (5,4)
10 Short essay produced by pair about primate (5)
11 Girl's article on small food shop (5)
12 Saintly child from Stowe accepting pounds to levitate mysteriously (6,3)
14 *The* Bond novel? (3,6,5)
17 Drop rest on creepy-crawly (4-4,6)
21 A rake might be more prudent around sandy desert state (9)
23 I may go after this character (5)
24 Pasty leading journalist declined (5)
25 Quiet, living in a place such as Cleveland? (9)
26 What a caddy should know: for example, a cart that needs pulling round (7)
27 No end to *Coronation Street*, soap shown regularly without a break (3-4)

Down

1 In agreement with forward on team (6)
2 Bread from dining room brought over after tea (7)
3 Valued a piano, much admired (9)
4 Betray twin on trial (6-5)
5 Spot or mole? (3)
6 Horrify a European after revolution (5)
7 Full? Ordered peer allowed in! (7)
8 Circumspect about river worker giving guarantee (8)
13 One qualified to handle a damaged plane? (4,7)
15 Dancing to it, again causing disquiet (9)
16 Film star, no sailor, on tug on holiday (8)
18 Former social worker without equal, covering miles on runs (7)
19 Ask earnestly to go in and dine, being short of energy (7)
20 Stop talking in jail (4,2)
22 Girl Irish boy stood up (5)
25 Name of writer, somewhat onomatopoeic (3)

18

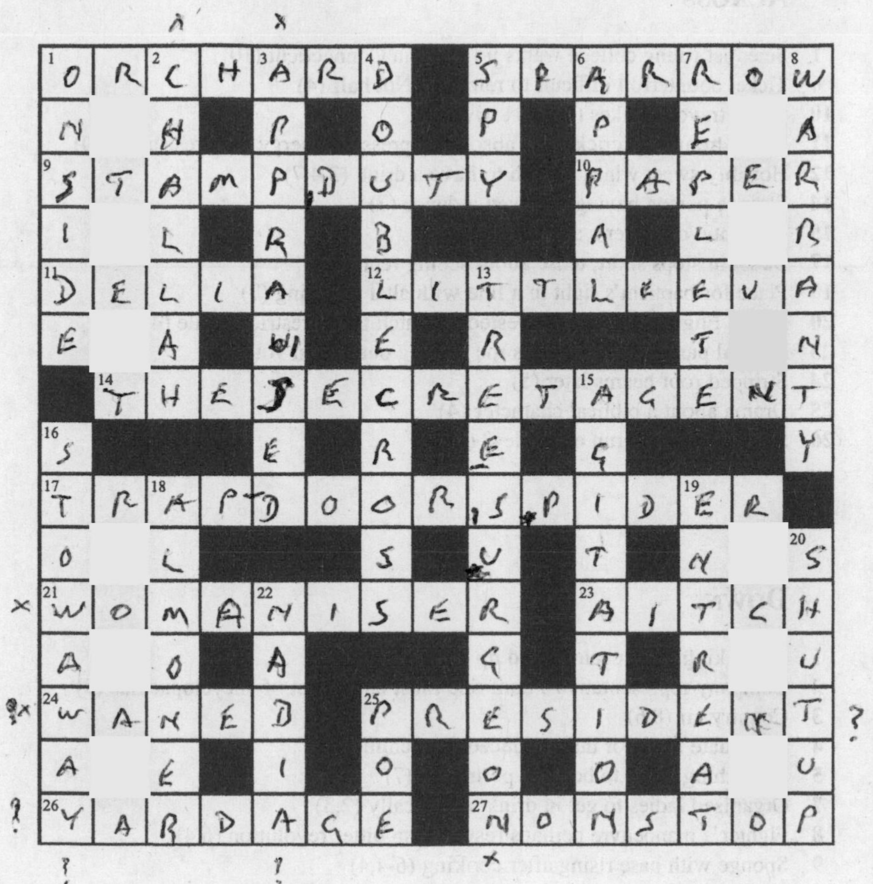

19

ACROSS

1 Scientist ruling college wants no artist in Connecticut (10)
6 Ticket counterfoil difficult to remove? Not half (4)
10 Space traveller takes time to arrive (5)
11 Inner strength of cricket bat absorbing pressure energy with resistance (9)
12 Horribly sweet wines — loth to have a drink (3,4,7)
14 Strange person having not even a dance (7)
15 Romantic daughter's a boring thing (7)
17 Bedouin stops short, tense about seeing reptile (7)
19 Place for baptism's right in a line with altar covering (7)
20 Public English company invested in watch for unrestricted sale (4-3-7)
23 Radical party's policy merits me turning out to pen vote (9)
24 Stripped roof beams later (5)
25 Drama about a biblical character (4)
26 Sport where feet run on blades? (4,6)

DOWN

1 Chess king and queen moved up for a pin (4)
2 Company representative's aim: one must have a set of encyclopaedias (9)
3 Country air (8,6)
4 Intimidate inventor that produced neat chime (7)
5 Everything's due to become permitted (7)
7 Organised ladies to get in drink specifically (2,3)
8 Fighter's manoeuvre perhaps resulting in bitter revolution (6,4)
9 Sponge with base rising after cooking (6-4,4)
13 Ponder most weird style of art (4-6)
16 Antiquated Missouri playhouse losing right to name (4-5)
18 Where are Maine, Rhode Island, California? (7)
19 Divers FBI agents chasing fine gold that's turned up (7)
21 One playing small part not included (5)
22 Title for knight has king elevating blade (4)

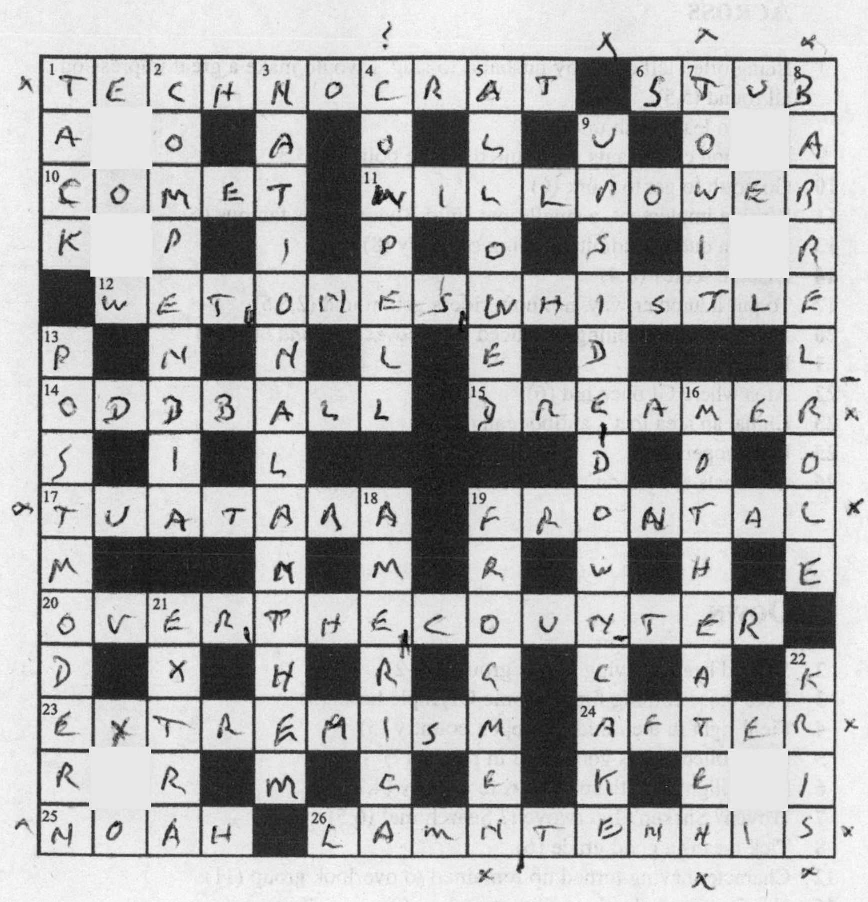

20

Across

1 Being often left alone by husband to sing, I would make a great impression all round (5,5)
7 Wine to leave with uncle (4)
9 Marathon contestants are wanted by the police (2,3,3)
10 Go north to get to work (6)
11 Writing implement, a small one: Enid Blyton's was famous (6)
13 Foreign queen's admitted being ordinary (8)
14 Dispute score? (8,4)
17 To put it another way, not how riders get around (2,5,5)
20 Inescapable happening produced by *deus ex machina*? (3,2,3)
21 Keep a piece (6)
22 Area where GI once hid (6)
23 China, an area that's antipodean (8)
25 Run properly (4)
26 Criminals as cartoon's disorderly characters (4,6)

Down

2 Second bean showing above ground (6-2)
3 Note those coming first in some Olympic heats (3)
4 Field right in the middle of open country (5)
5 Debt collector has good time in prison (7)
6 Feel delighted with small part to display (4,2,3)
7 Woven? Shaken? Interwoven? Search me! (6,5)
8 Tick means a good grade (6)
12 Character having turned up remained to overlook group (11)
15 No cigar in production without carbon (9)
16 No orthodox believer in free trade with oil (8)
18 Ghastly retreats after time ran out (7)
19 17 pronouns in system (6)
21 26's leader given name "Chicken" (5)
24 Something police do in southern states (3)

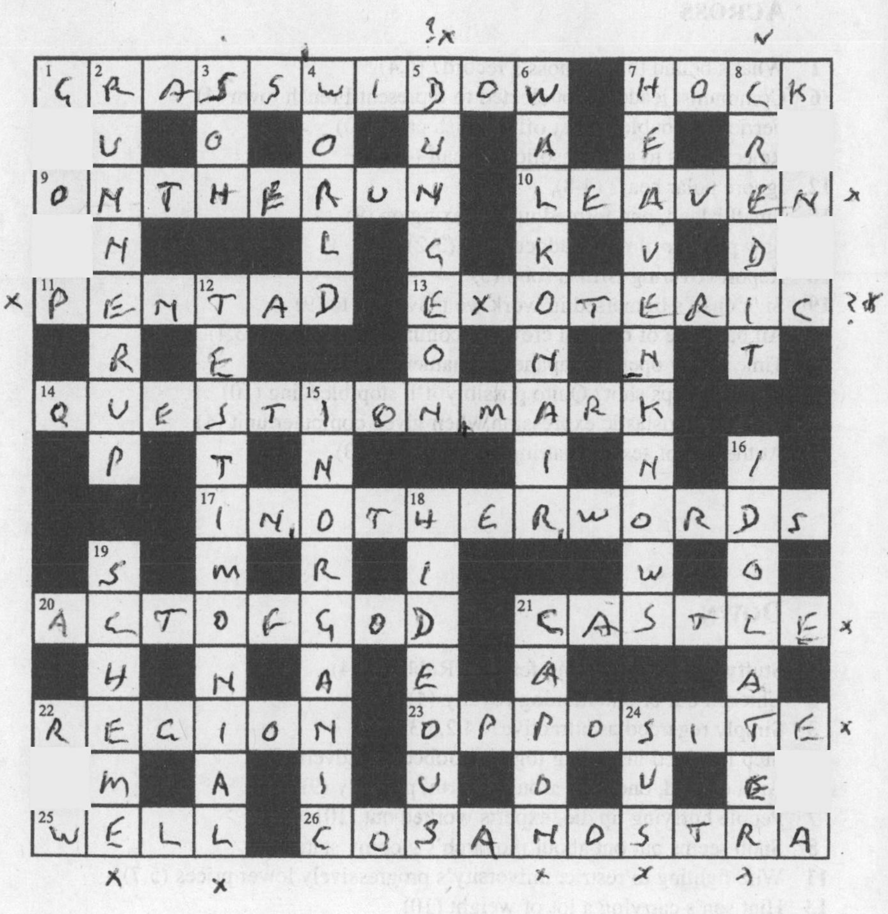

21

Across

1 What's bound to hold boss's record? (6,4)
6 Communist leader's not elected to represent French town (4)
9 Ferries in trouble at first off Spanish cape (10)
10 Reject plans to send unsolicited mail (4)
12 Ignore polar bear? (4-8)
15 Small island, one formed up in Lexington (9)
17 Saw popular film clip added later (3-2)
18 Report covering British food (5)
19 Sign chap's immersed in work we may refer to (9)
20 All bar none of oriental crew welcoming jolly sailor (5,3,4)
24 Time lost in opening Japanese ornamental box (4)
25 Nothing stops shot? Quite possibly it'll stop bleeding (10)
26 Nerd's enthusiastic expression when given computer unit (4)
27 At the end of session batting with partner (10)

Down

1 Stuffy air over admitting female? Rubbish! (4)
2 Chicken out of entertaining royalty (4)
3 Simply regarded as attractive? (4,2,3,3)
4 Step involved in putting together Jobcentre advert (5)
5 Area of land, one seen around North, possibly (9)
7 People hurrying up die, experts worked out (10)
8 Snob seems put out about monarch's gloomy state (10)
11 Wife fighting to restrict university's progressively lower prices (5,7)
13 Hint son's carrying a lot of weight (10)
14 Hear aristocrat's to join county club (10)
16 Gathered somebody seen outside was in possession of drug (9)
21 Judge heading off from court can hold a lot of drink (5)
22 Point for Scots to ponder (4)
23 Wait for a guy (4)

22

Across

1 Sick feeling: good to be in house (8)
9 In river, an explosive that's being tested (8)
10 Fighting force cutting policeman in two? (4)
11 Careless girl has small child briefly to look after (6,2-4)
13 Southern European held up as fine (6)
14 Butcher's success: selective slaughter done for the cooking (8)
15 Shame one's ruined well-loved music (7)
16 Note-writer has to alight extremely swiftly (7)
20 Number visiting doctor in turn, having a bit of a fall (8)
22 Not all lost at a mission that's under the Japanese (6)
23 Full of energy, have success with serious opera (4,8)
25 Idle talk in British paper (4)
26 Flyer may have accident at track (8)
27 Religious practice, but wrong start in mass (8)

Down

2 Coil in a river that's grown (8)
3 Courtier's end resulting in chaos (12)
4 Reckon priest came over to some soldiers to give a bit of backbone (8)
5 Substitute coarse material with fine (7)
6 Spring imprisoned doctor (6)
7 Joint may present problem to maid (4)
8 Tired, keeping pressure on arms (8)
12 Settle down in Inn: cheeky lad comes in for lawyers (4,2,3,3)
15 Town will have second university in a few weeks (8)
17 Vladimir's fellow-waiter is French, badly-dressed? (8)
18 Near miss suffered — the discussions are academic (8)
19 Be likely to catch fish, coming up in this? (4-3)
21 Bachelor almost entirely keeping free of marriage (6)
24 Superior companion appears spooky after the start (4)

22

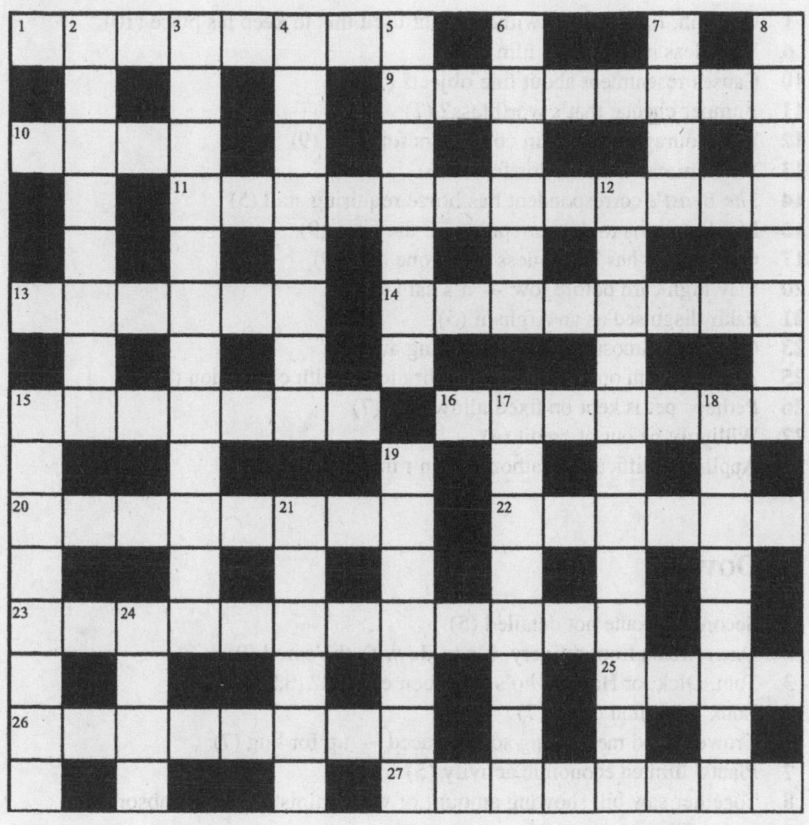

23

Across

1 Obadiah, for example, without right used this to keep his place (10)
6 Worthless man in dirty film (4)
10 Causes resentment about fine objects (7)
11 Bumper cheque that's worthless? (7)
12 Is discouraging about tin containers for wine (9)
13 Wild animal heard in this forest (5)
14 *The Beast's* correspondent has horse requiring stall (5)
15 Monkey starts to devour spiders in the trees (9)
17 Old country has bottomless pit in one area (9)
20 Play high card before low — it's safe (5)
21 Fakir disguised as an Afghani (5)
23 Giant fish almost repeated, shocking aunt (9)
25 A doctor with one medical speciality to do with circulation (7)
26 Perhaps pet is kept on fixed allowance (7)
27 Willingly be out of credit (4)
28 Appliance difficult to fathom? Don't move! (4-6)

Down

1 Secondary route not detailed (5)
2 One retreats from refinery; it's to do with the smell (9)
3 Tom, Dick, or Harry, who's just been evicted? (3,2,3,6)
4 Look up to find detail (7)
5 Crowd round me, getting so embraced — up for hug (7)
7 Plants' limited economic activity (5)
8 Together saw bill showing amount of work minister's house absorbs to run (9)
9 One proverbially keen on column I dropped is put on for soothing (7,7)
14 Boycott gloomy dance (9)
16 Ignoring visitors, barking at the moon (3,2,4)
18 Get GI in, the worse for wear (7)
19 Stores going down the chute? (7)
22 Fine and fit? It's not true (5)
24 Division in church? I'll say! (5)

ACROSS

ACROSS

1 Agent snatching short sleep without hesitation (6)
4 Very narrow opening for hopeful travel operator (8)
9 Someone unable to comprehend an erroneous description of Pandora's box? (2-5)
11 Match spreading fire? Reduced chance up front (7)
12 Group of experts criticise Spanish article (5)
13 Sharp or flat? That's not normal (9)
14 With issue to raise, one's in such a state (10)
16 Fine performance is a matter of record (4)
19 Crowd not beginning to hurry (4)
20 Not having benefited from the class system? (10)
22 Drug flow leads to remarkable increase in dried leaves (3-6)
23 Suggest just refusing to start (5)
25 Shakespearean setting badly affected by atmosphere, on reflection (7)
26 Black out film extracts about drug repeatedly? On the contrary (7)
27 Teams straddling path laterally (8)
28 Inn servant had role to play over the way (6)

DOWN

1 Abrasive extremes from sharp impressionist (9)
2 Pale woman featuring in article (5)
3 From oil transporter, spot East River rising (8)
5 Ultimately exhausted, getting personal best in race, say (3,2,4,4)
6 Read out liturgy around city (6)
7 Currently successful soldier is source of dope (9)
8 Lots of old information turning up about name in register (5)
10 Approximate description of hooligan with money? (5-3-5)
15 Alien, amidst others, went ahead and found new home (9)
17 Newly-made bed ready? Time to go up — with this? (5,4)
18 Old fighter pilots maintaining height (8)
21 Make use of library for writer of gypsy book (6)
22 Golfing standards involving one in golfing matches (5)
24 I learn part of my anatomy (5)

24

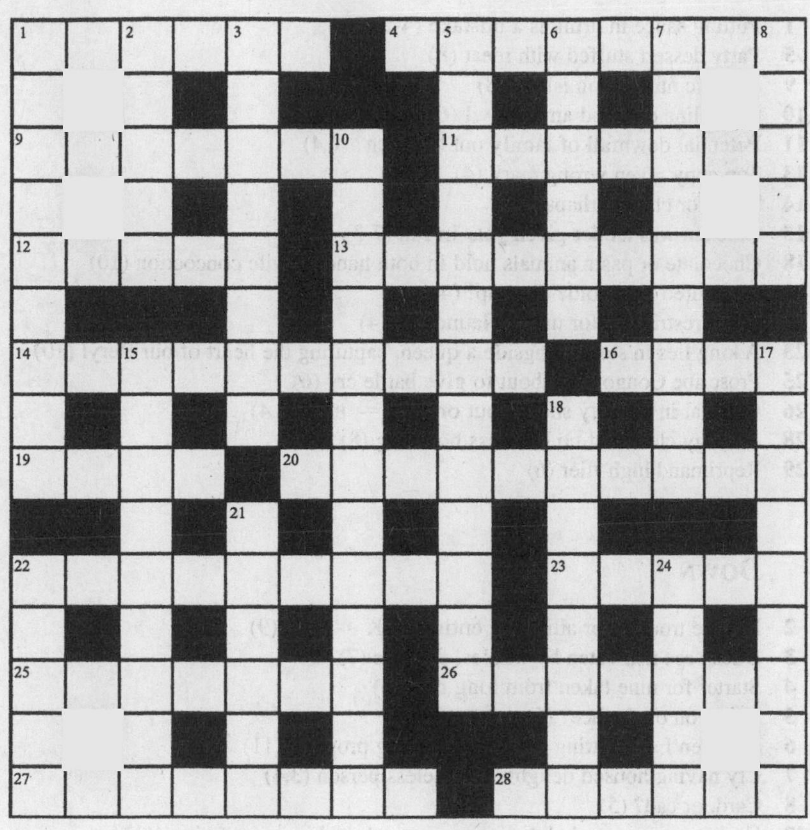

25

Across

1 Putting sauce in drink is a mistake (4-2)
5 Party dessert stuffed with meat (8)
9 Specific number on island (8)
10 Cue a line on good among evil (6)
11 Potential downfall of family out to lunch? (6,4)
13 Top copy given wrong mark (4)
14 Time for chat, perhaps? (4)
15 Chicken was tender given year in run (7-3)
18 Chocolate or pasta animals held in both hands in vile concoction (10)
20 A hundred years old? Shut up! (4)
21 Local restriction for missile launcher? (4)
23 A king lies in state alongside a queen, capturing the heart of our Beryl (10)
25 Proscribe Congo, not about to give battle cry (6)
26 Oriental in dynasty shortly put on food — this? (4,4)
28 Strategy clear, but no business booming (8)
29 Reprimand high-flier (6)

Down

2 Double trouble for allies not entirely OK — OK? (9)
3 Duck sees fish eaten by whales, perhaps (7)
4 Starter for nine taken from long dish (3)
5 Nerve on one's face? (5)
6 Batsmen I see getting out, not attending properly (11)
7 Cry having housed delighted homeless person (3,4)
8 Card, or cad? (5)
12 Oasis touring round clubs set up musical, making connection (11)
16 Tool gone missing, not old (3)
17 South American barman maintaining man an upright character (9)
19 One highly-placed calling for service (7)
20 America nearly mad, about to become potty? (7)
22 Quickest way out of cold, cold lake? (5)
24 Posh Conservative grabbed by maniac, rough (5)
27 Blade beheads pig (3)

25

26

Across

1 Wicked foes ruin a manoeuvre (9)
6 Unscrupulous character on the staff (5)
9 Gander walks unevenly in geese's enclosures (7)
10 Complaint made as food's left in a slightly different place (7)
11 Forum for discussing some heresy, no doubt (5)
12 Melancholy individual accepts disturbance in radio frequency (9)
13 Bar serving drink to actor around ten (4-4)
14 Shabby-sounding carriage (4)
17 Excursion often occurring just before the fall (4)
18 Run-of-the-mill doctor covering old part of army (8)
21 Pork pie supplied by poet with quarters on river (9)
22 Case a woman will briefly make (5)
24 Agreement to clothe soldier in ribbed fabric (7)
25 Lose shine — a bit like a mountain lake? (7)
26 Dangerous female leaves in high spirits (5)
27 American celebrity deceived about English vessel (9)

Down

1 Hot drink, say, imbibed by rising star (5)
2 Once Salote's state of detached amicability? (8,7)
3 Shame about girl joining artist's expedition (8)
4 One produces fruit in traditional short feast (8)
5 Description of Orion given by leading player on line (6)
6 Shrubby tree: pig ate the lot (6)
7 Info categorised wrongly — now I'm responsible (3,2,10)
8 Favourite weapon restraining universal ill humour (9)
13 Deception beginning to repel army mechanic (9)
15 Feud cleric had with one he should have forgiven, say? (8)
16 Strolling player, the last person to inspire a lay! (8)
19 Where Caesar was overthrown, in yours truly's recollection (6)
20 It may keep one on top of the jumbo! (6)
23 Writer holding horse for composer (5)

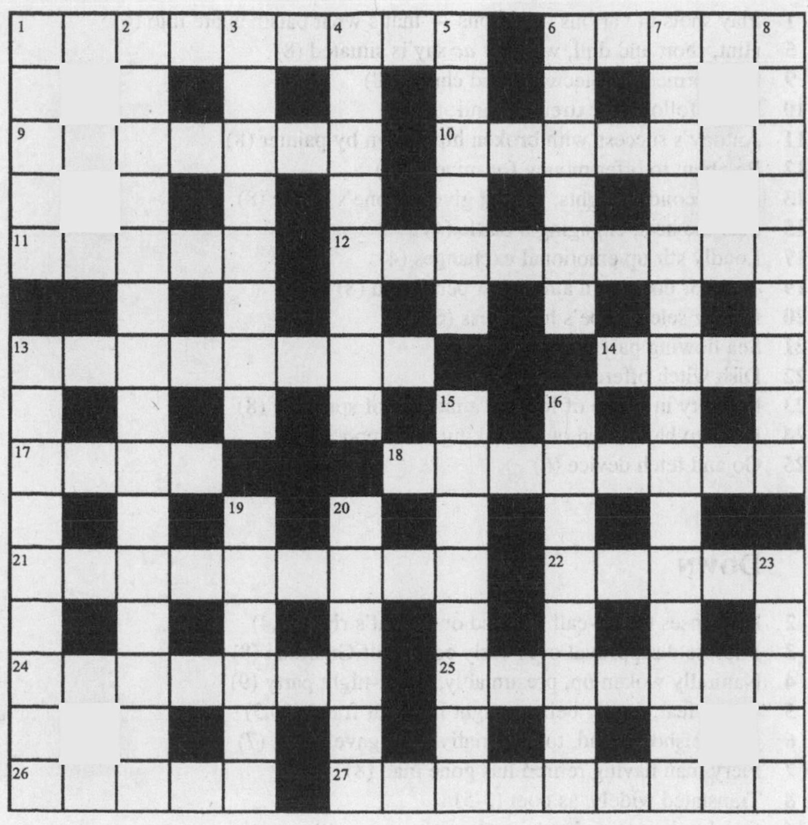

27

ACROSS

1 Play shots in various directions — that's what batsmen are into (6)
5 Hint, short and dull, where 1 *ac* say is situated (8)
9 God formerly protected round church (8)
10 Line I followed, extremely uniform (6)
11 Antony's success with broken hip shown by painter (8)
12 Be about to offer money for mineral (6)
13 Had second thoughts, having given in one's notice (8)
15 Feel the heat, engaging a battle (4)
17 Loudly stir up emotional exchanges (4)
19 Team of doctors in attempt at perfection (8)
20 Finally select pope's headdress (6)
21 Sea flowing past bridge gap (4,4)
22 Dish witch offered to soldiers (6)
23 Left city in hands of fool, in a manner of speaking (8)
24 Right to have cried out, breaking fine Spode (8)
25 Go and fetch device (6)

DOWN

2 Responses to roll-call shouted out? That's right (4,4)
3 Express disapproval over early edition of Guardian (8)
4 Naturally woken up, presumably, by all-night party (9)
5 Secret fear, finally being caught in credit fraud (10,5)
6 Clear bishop ahead, took initiative and gave check (7)
7 Fiery man having retired has gone mad (8)
8 Translated widely, as poet (3-5)
14 Good to be upset about murder: it's wrong (9)
15 Playwright's line saved by collector (8)
16 Old woman stayed in, having got mean (8)
17 Irritated about English business having shrunk (8)
18 A know-all on the water, I take new crew out (8)
19 In worse physical condition, being comparatively mountainous? (7)

27

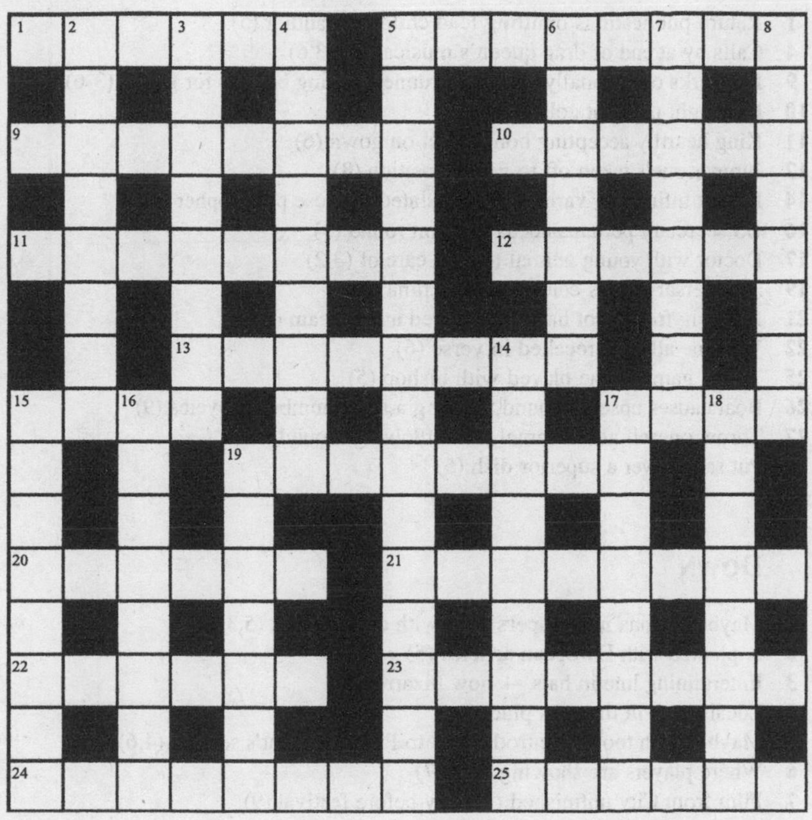

Across

1 Return publications omitting lead children's author (5)
4 Calls by at end of drag queen's musical act (3,6)
9 He works occasionally with rum runner, getting billions for goods (3-6)
10 First light paper at college (5)
11 King heartily accepting honour put on gown (6)
12 Jumper one's taken off to get in position (8)
14 Insight ultimately variable in translated Chinese philosopher (9)
16 US secretary perhaps needs weapon round (5)
17 Doctor with young animal to take care of (3,2)
19 Anniversary that's celebrated in China (9)
21 Building free of rot having removed a fine beam (4-4)
22 Wartime alliance recalled in verse (6)
25 Tricky game Rome played with bishop (5)
26 Boat causes upset in sound, passing a large number of cycles (9)
27 Throw oneself about camel, sand blowing around (4-5)
28 Put sauce over a superior dish (5)

Down

1 Maybe Havana newspapers hold with deception ... (5,3,7)
2 ... planted with European articles (5)
3 Entertaining line in hats — how bizarre! (7)
4 Locals born in timeless places (4)
5 Maybe Noah took on introduction to Prince Regent's servant (4,6)
6 Where players are showing guts (7)
7 Film from City unfinished one day before festival (9)
8 Wrongly put frozen heart before universal literary hero (6,2,7)
13 As sprinklers do, accelerate with fire spreading (10)
15 Covering top of bush, a high plant (9)
18 Paid the rent for an area of mine (7)
20 Journeys to find modern music (4,3)
23 Woman on her first kid (5)
24 One good look, eyeing their tops? (4)

28

Across

1 Hindu retreat remains next to sheep (6)
5 Losing team's disadvantage (8)
9 One into rotten stuff, or one in the printing industry (10)
10 A boundary retrieved proves turning point (4)
11 Hunk admits vice naughty (8)
12 Piece for Frenchman, *The Marseillaise* perhaps (6)
13 In which are capitals of Azerbaijan, Singapore, India and Afghanistan (4)
15 Sock won't begin to be stuffed in backward Christmas — formal occasion (8)
18 Done in posh way, Queen enters sea (8)
19 Audibly wretched manner (4)
21 Bowler, possibly, gets out (3,3)
23 Measure in pot a Bulgarian's money (8)
25 Broadcast includes hospital programme (4)
26 Short of time, got pelican running wild around end of day — hard bird to shoot? (4,6)
27 Very much Conservative rule, but surprisingly European (4,4)
28 In solid fashion? Not at all! (6)

Down

2 Shilling cut, point rise (5)
3 About to get what comes before proper talking-to (9)
4 Army providing range (6)
5 Studied case for corruption involving leadership — a growing problem (5,3,7)
6 Pervert isn't commonly given make-up (3,5)
7 Sting turned out nicely (5)
8 Stress in evening rush hour? (5,4)
14 Someone whose boots filled by Hercules' labouring? Without question (9)
16 Dim failure gripped by desire becomes excellent specimen (9)
17 Expert to pass code (8)
20 Welly filled with a metre of brass (6)
22 Part of the bow was untidy, occupying man (5)
24 Clever-clogs ignoring Washington achieves minor eminence (5)

30

Across

1 Nominal source of condiment (10)
6 Fail to include it with high honour (4)
9 Heavy flow of water from fissure behind hill (7)
10 Drinks spelling the end for flirting girls (7)
12 Unskilful writer held back in computer studies (5)
13 Absolutely false story about end of tight road race (9)
14 Term for a night one gets drunk, nothing less? No, somewhat later! (3,7,5)
17 It may leave you a shade disadvantaged (6-9)
20 Against two blokes being found guilty (9)
21 Showing pace and power in attack (5)
23 Friend taking in fashionable European theatre performance (7)
24 Allowed English sailors to snatch key item of church furniture (7)
25 Vat smelt? Don't open it! (4)
26 A complete catalogue of surfing venues? (10)

Down

1 Description of a scoundrel when the chips are down? (9)
2 Good to engage in absolute clear out (5)
3 Court involved in vote? Shocking (13)
4 A grateful expression maintained by family from part of Spain (7)
5 Trendy seaside resort (not British) (5-2)
7 Berries suitable for those feeling peckish? (9)
8 Attractive way to dive into river (5)
11 Broadcasts with grand sporting experts affected attitude (4,3,6)
15 Leader of band heading off leader of soldiers (9)
16 Home refrigerator ultimately needs ice shifting (9)
18 Source of charge — assault? (7)
19 Favour fancy duelling, dispatching one learner (7)
20 Wake up, having missed last astronomical sight (5)
22 Devotional painting — some sweet expression of thanks (5)

30

31

Across

1 Proclaim and declaim in the chair (12)
9 Biological expert is a follower of Democritus (9)
10 Lord, there's nothing harvested, it seems! (5)
11 After short time see animal move unsteadily (6)
12 Volunteers housed in farm building, as may be disclosed (8)
13 Sleep needed by little man — endless sleep after being here? (6)
15 Maniac finally found out went to pieces (8)
18 Flowering plant brought in by radiant husband (8)
19 Rare shock when Conservative gets in (6)
21 Retaliating fighters spraying nerve gas (8)
23 I'll take English and do it differently, right? (6)
26 Child's vehicle is hit — front comes off (5)
27 Plant girl exposed to the wind (9)
28 Statement perhaps from Holy Writ (3,9)

Down

1 Carriage from Westminster heading north to a school (7)
2 Love some abstract works at Tate Modern? (2,3)
3 Particular group of people without work making a sort of escape (9)
4 Endless muck to make one lose weight (4)
5 Associated with spinning rubbish — a politician (8)
6 Country bird trapped live (5)
7 Footballer who acts like a baby? (8)
8 Plan to be at home having nurse (6)
14 Gradually introduced new shade in 18 *ac's* colour, almost (6,2)
16 A comedian out in the country (9)
17 Completely unfashionable, favouring conservative politics (8)
18 Shortage of land to support duke (6)
20 Serious article penned by Hemingway (7)
22 Great delight when bishop enters church territory (5)
24 The competition — he avoided draw (5)
25 Endless luxury brings advantage (4)

31

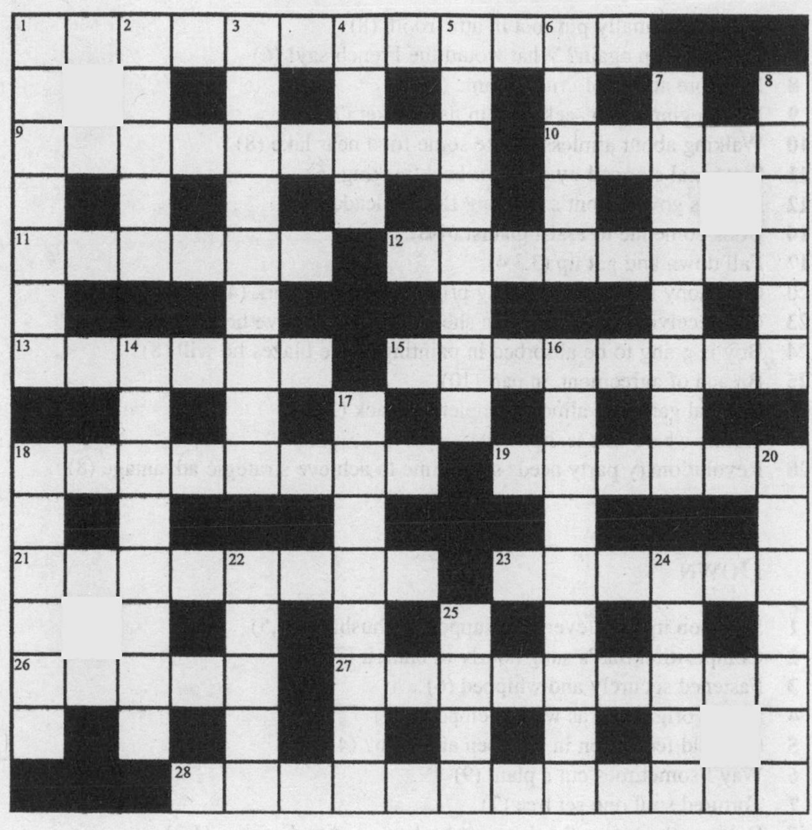

32

ACROSS

1 Mate, see, finally put foot in attic room (8)
5 Lobster soup again? What would the French say! (6)
8 No score at end of crucial game (3)
9 Bird beginning to seek food in fish basket (7,3)
10 Walking about aimlessly, take some food near lake (8)
11 State trial stopped by emperor backtracking (6)
12 Regrets government's ignoring Liberal leader (4)
14 Book someone to assist pianist (4-6)
17 Fall down and get up (3,3,4)
20 Ceremony recognised retiring prime minister's work (4)
23 One receiving goods is given shelter by cove, we've heard (6)
24 Boy is going to be absorbed in painting? Like blazes he will (8)
25 Breach of agreement, in part (10)
26 Clerical garment, almost completely black (3)
27 Home: where this is, hot? (6)
28 Revolutionary party needs some time to achieve strategic advantage (8)

DOWN

1 Summon in court everyone supporting husband (4,5)
2 Composition that's sung hourly in church (7)
3 Fastened securely and whipped (6)
4 Friend originally, as well as emperor (9)
5 Get cold feet when in the open air again? (4,3)
6 Way I sometimes cut a plant (9)
7 Tortured soul one set free (7)
13 Reluctantly accept fleet's to get broken up at end of war (6,3)
15 Holy book's writing style showing tendency to digress (9)
16 Trustworthy delegate on university board (9)
18 Conceive a slogan for anti-Ecstasy protester (7)
19 Two types of wood found beneath bottom of pile? Nonsense (7)
21 Undertook, in a war, to capture island out east (7)
22 German city that is attractive to northerners (6)

32

33

Across

1 Post application at once, I'd say (5,3)
6 Blue needed by country missing a game (6)
9 Tasks one arriving with capture of stronghold: no relief from this source (4,9)
10 French author working as judge (6)
11 Artist impossible to resist (8)
13 Town named after historian? Daughter's confused (10)
15 Magwitch gives a short signal (4)
16 People generally addressed in the country (4)
18 Sackville-West taken back in charge in case (10)
21 Whip is available in white (8)
22 Area of dry hill (6)
23 Sacred island, note, beyond range? That's pitiful (13)
25 Babywear put on son: make it fast, so (6)
26 Remedy is neat solution covering one spot (8)

Down

2 Delight about author reciting (7)
3 Game played by two of Turandot's ministers? (5-6)
4 Nothing to read in this language (5)
5 Iron, good in the bone (7)
6 Several Edmunds in Cromwell's army (9)
7 Almost excessively be fond of girl (3)
8 Kill in furious rage, too much to suppress (7)
12 Stretcher! It's clean, bad break (7,4)
14 Fancied being in a line, receiving bearers of gifts (9)
17 Complain about this Roman elevated style (7)
19 During tea, different sort of bun offered (7)
20 Extreme north shown in colour (7)
22 Take illegal drug and a little whisky (5)
24 Mark a page in guide (3)

33

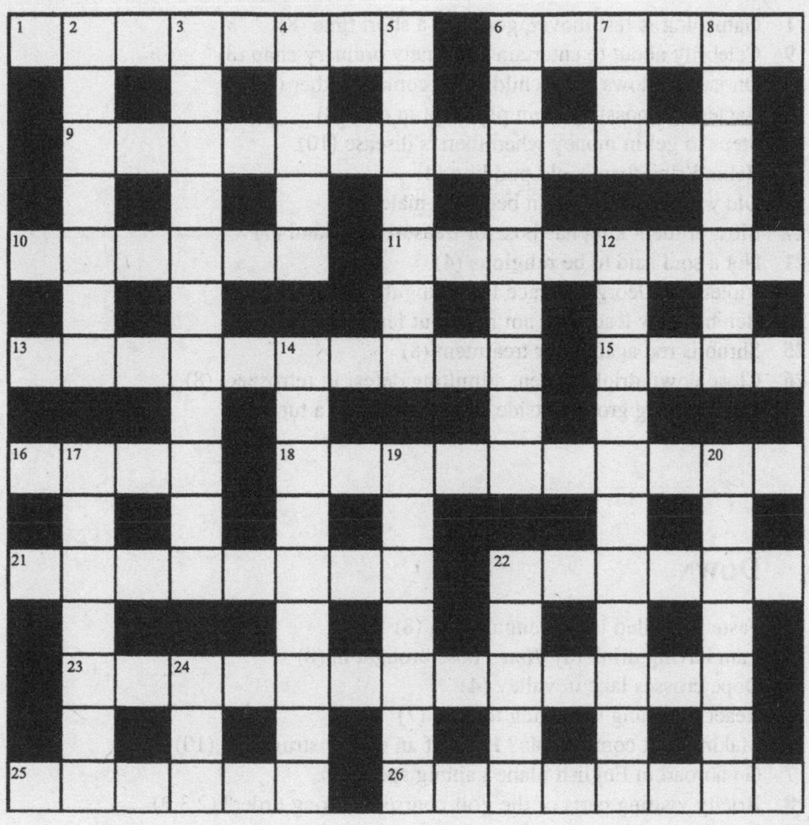

Across

1 Game dog is fast mover, good for a short time (8)
9 Celebrity about to entertain extremely ordinary chap (8)
10 On jaunt, allows three children to come together (8)
11 Bacterium, possibly from pig, kept in dish (8)
12 Steps to get in money when there's disease (10)
14 Behind sink there's old pudding (4)
15 Old woman cavorting in bed with male (7)
17 Most affluent king has box for treasure on island (7)
21 Not a soul said to be religious (4)
22 Violet and Georgia in race following the route (10)
23 Ben has new leader — not male but female gusher (8)
25 Shrub is red again after treatment (8)
26 Close down drinking den, admitting defeat in retrospect (8)
27 Broadcasting group outside university doing a turn (8)

Down

2 Fastener pulled in opening curtain (8)
3 I am having drink my *Times* boss brought in (8)
4 Dope crosses lake in valley (4)
5 React, touching son going to pool (7)
6 Making nest comfortable? Hard, if an egret's struggling (10)
7 Go abroad in English plane gaining speed (8)
8 Briefly visiting parts of the golf course in wrong order? (2,3,3)
13 Feeble new board is awful (10)
15 Genuine bind: a foe being nasty (4,4)
16 Left facing pain, grow weak (8)
18 Shell may be one of those lined up for this business building (8)
19 Poisonous creature on crops I zapped (8)
20 Craven gerrymandering restricts one seeking to get even (7)
24 Test shows sea creature lacks carbon (4)

34

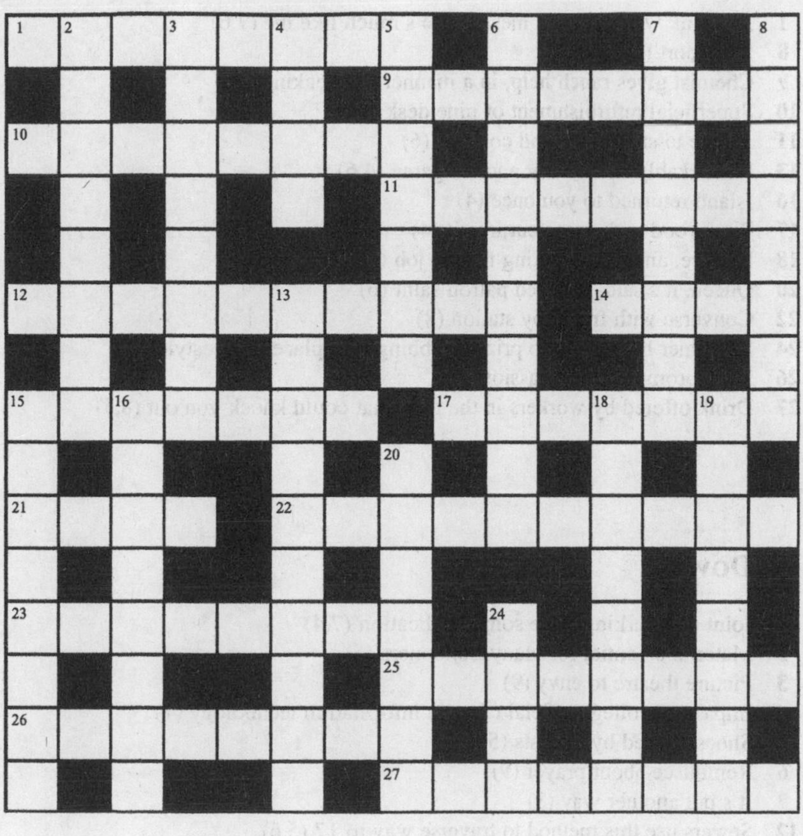

35

ACROSS

1 I'd drink with priest in mess — he's much like me (7,6)
8 Shell port (4)
9 Chemist gives ranch help, in a manner of speaking (10)
10 Superficial refurbishment of pine desk (4-4)
11 A note to strengthen and comfort (6)
13 Remarkably silent, near ancient game (4,6)
16 Island returned to you once (4)
17 Pine wood with name cut into it (4)
18 Lecture, and form a liking for the job (4,2,4)
20 Queen, it's said, pictured patron saint (6)
22 Converse with friend by station (8)
24 Swimmer has a right to prize, grabbing first place in freestyle (10)
26 Babe prompting compassion (4)
27 Drink offered by workers in the field that could knock you out (8,5)

DOWN

1 Joint not working? Use some application (7,4)
2 Material essential for many Londoners (5)
3 Picture theatre to envy (9)
4 Impassive college official takes in information technology (7)
5 Shoes needed by cyclists (5)
6 Reminisce about prayer (9)
7 It's put another way (3)
12 Sewers use this method to traverse way to 17 (5,6)
14 Symbol king applied to goddess, showing excessive garrulity (9)
15 The pup's so determined to start operations (3,2,4)
19 More nuts fell, right? That's right! (7)
21 Best? Yes and no (5)
23 Vamp's passion splitting partners (5)
25 Tear, as journey can't start (3)

36

Across

1 Reportedly a Mediterranean island's symbol of mourning (7)
5 He died, after retirement; son wants satisfactory biographer (7)
9 Odious English attorney old Weller engaged, dividing payment (9)
10 Tree, in the role of writer (5)
11 Soldier bodily put on weapon (5-8)
13 Scrape involving supporter in a Swiss town (8)
15 Extravagant priestess in charge (6)
17 Vivacious wit brings imp's energy to the fore (6)
19 Plot originally conceived on shoot (8)
22 Be on the ball and able to identify Scottish engineer, say (4,5,4)
25 Saw Lady Macbeth's poor cat mentioned here? (5)
26 How *Lampyris noctiluca* shines, with great enthusiasm! (9)
27 Journeyman observed thousands in woody plant by river (7)
28 Engraver's house with courtyard surrounded by cloisters (7)

Down

1 Eccentric king or queen, for example (4)
2 Union leader in East London appealing to the public (7)
3 It could be Terry, the Lady of the Lake (5)
4 Outline of ode finally penned by court poet (8)
5 Briefly, one doubting graduate's first descent into triviality (6)
6 Being quick to perceive faulty intonation (9)
7 Initiator of escape thus accepts role as grass (7)
8 Country philosopher died cut off from the sea (10)
12 Rare task, two universities developing German course (10)
14 Calamity that may cost someone a packet! (9)
16 Indeed true: duck is found in river (8)
18 Academic has a new drug put to improper use (7)
20 A number in part of army understand set-up (7)
21 Aircraft accommodation no good in sea-mist (6)
23 Moving fast, mercury flows round one in Scotland (5)
24 Autonomy that's not entirely a false notion (4)

36

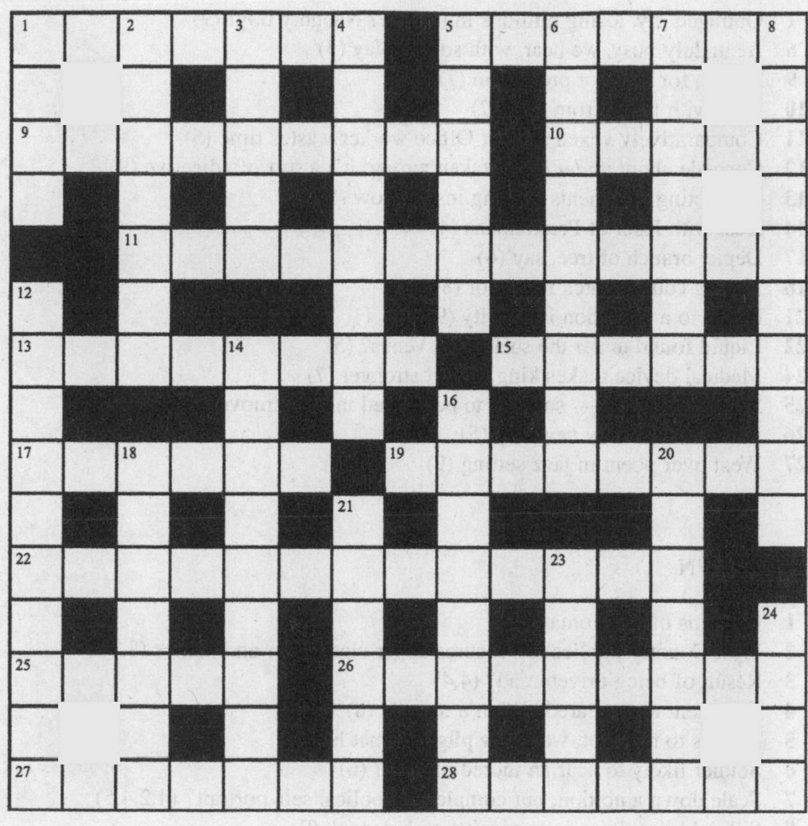

37

Across

1 Outraged cry, losing millions in fortune? Naughty boy! (9)
6 Be unduly busy, we hear, with sort of play (5)
9 Fleece, for one, for protection (7)
10 Help with froth, from her? (7)
11 Comparatively vexed as Post Office worker wastes time (5)
12 Concede about *under* being taken wrong: it's a sort of adjective (9)
13 Distracting opponents playing in the snow (8)
14 Deal with loser of Prestonpans (4)
17 Depict branch of tree, say (4)
18 Having cause, watch regulator (8)
21 Points to a reduction in gravity (9)
22 Liquid found under the surface of Venus? (5)
24 Medical device makes king's heart stronger (7)
25 How paras go in — soldiers to be backed in rapid movement (7)
26 Rejected house by new city (5)
27 Went over poem in jazz setting (9)

Down

1 Customs of St Thomas's (5)
2 West Country publication secures upper-class actor and author (8,7)
3 Result of being driven mad? (4,4)
4 Opponent disappeared? What a tragedy (8)
5 Thanks to the poet, we know pilgrims met here (6)
6 Soldier likely to hear an incredible tale? (6)
7 Scale down ambition, but complete diabolical self-portrait? (4,2,4,5)
8 Silly old people go in to receive cash deposit (9)
13 This army so valiant, getting involved? (9)
15 State was to ban criminal (8)
16 High-class sports science qualification? I came fourth (8)
19 Sensible lens regularly used by film director (6)
20 Feel sorry, having run over bird (6)
23 Quick attack nets pawn (5)

1
2
3
4
5
6
7
8
9
10
11
12
13
14
15
16
17
18
19
20
21
22
23
24
25
26
27

38

ACROSS

1 Providing revolutionary, bodies trapping copper fight (10)
6 Course it's a nail (4)
8 Dead without a flower, one given wreath (8)
9 Attractive girl round piano creates a bit of competition (3,3)
10 Hostile introduction also called an intimidatory chant (4)
11 What might be cunning, as eats mice? (7,3)
12 Where caps found in one instance on part of shoe (5,4)
14 Trainee officer needs time to follow rebellious Jack (5)
17 Innocent went back to start of tunnel, to break out (5)
19 To summon youth leader in spoils reprobate (9)
22 Platform, unknown length, wants simple floral arrangement (5,5)
23 Dry river — save its environment (4)
24 Run through with stake, I appear anaemic (6)
25 Fake Titian taken in that is found (8)
26 Turner to show contempt (4)
27 Trouble before entering service — she'll take charge (10)

DOWN

1 Promising hand with well-attended party (4,5)
2 Pretentious food in first half of meal (5-2)
3 Alas whirling into tilt, nebula of mine? (4,4)
4 Psychoanalytical test I recite on sofa, as ordered (4,11)
5 Fool given help, we hear (6)
6 Truck goes over a track, a long way to go (9)
7 Say what mountaineer has to do to mountain, given weather (7)
13 Revolt of peasants restricting chairman's last term on board? (2,7)
15 Hardy character admitting bad night with inebriation (9)
16 Damned response to grit in eye? (8)
18 Plan these days in order to break strike (4,3)
20 Struggle giving up unprocessed food (7)
21 Get lost round end of lane — cry loudly (6)

38

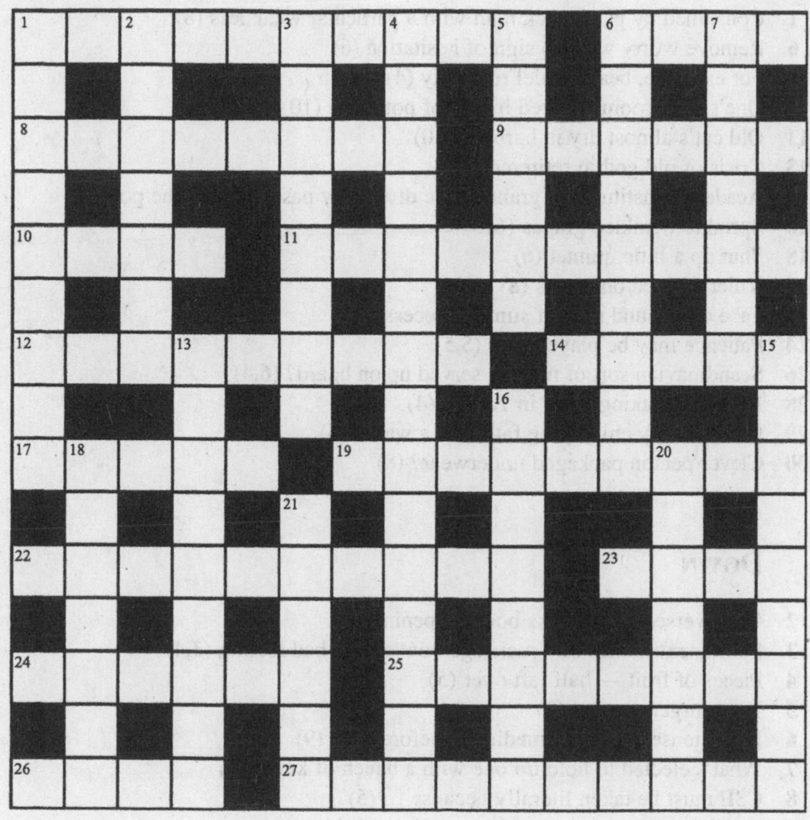

39

Across

1 Consumed by pains, sick man with a particular weakness (8)
6 Remove worry with no sign of hesitation (6)
9 For example, beat spaniel regularly (4)
10 One's little room covered in lots of potpourri (10)
11 Old cat's almost dry in harbour (10)
13 Crisis of old god in retirement (4)
14 Academic institution's grand noble drunkenly passed round the port (8)
16 Sprinkle wrinkled grapes (6)
18 Shut up a little quintet (6)
20 Killer entered on beasts (8)
22 Take off around start of summer recess (4)
24 Patience may be played in it (5,5)
26 Scandinavian sort of movies served up on board? (6,4)
28 Bold girl making mark in *Times*? (4)
29 One possibly employing father as a waiter (6)
30 Clever person packaged underwear? (8)

Down

2 One overseeing timeless book's opening (9)
3 Putting extra lines into marriage vow causes bad feeling (3,4)
4 Pieces of fruit — half left over (5)
5 Notes urgent appeal (3)
6 Break tenses found immediately before this? (9)
7 What's elected to hold up one with a bunch of keys? (7)
8 CSE must be taken literally because … (5)
12 … mean to get exam level lowered past it (4,3)
15 Who's nominated by the groom, new daughter being cradled? (9)
17 Slide between bars? (9)
19 Most of them succeeded without mine — that's bad (3,4)
21 Cold character slammed by *Woman's Own*, at last (7)
23 An instrument raised capital for the locals (5)
25 Palm about to be planted in region (5)
27 Put one over adversary in court? (3)

39

40

Across

1 Account in newspaper should be this (4)
3 Insurgent arm in quiet Caribbean spot (10)
9 Unusual articles out of the *Listener's* archive originally (7)
11 Male office staff touring heart of Turin (7)
12 Watch out for this sporting event (4,5)
13 It's absurd copying exercises in a line (5)
14 Traditional match win edged with confusion (5,7)
18 Disheartened lady in shopping centre sounded nervous (6,6)
21 Get round course finally, following chap back (5)
22 Stump sticks out of grass, reportedly, extremely loose (9)
24 Boost following express delivery (7)
25 Father and I drinking bad hock with cabbage (3-4)
26 Kid's heading for row, taking in stray dog (3-7)
27 Hang about with pack of Aussie travellers (4)

Down

1 Have spare time at first in African city (8)
2 Say farewell and head for some food (4-4)
4 Some acclamation for instrument-maker (5)
5 After collapse doctor put on variable spin, with hot air (6-3)
6 Rural town keen to develop computer system (6,7)
7 Parisian who upset clergyman gets a shake (6)
8 Boy or girl equally affected by name dropping (6)
10 Expected rank first held by churchwarden (2,3,8)
15 One hit boxer, shattering his display (9)
16 Mastermind maybe succeeded in grilling method (4,4)
17 At one time holding Bible classes in centre of Calgary (8)
19 Live on 50 per cent benefit (6)
20 After dull run football team climbs table (6)
23 Maiden climbs to top of large tree (5)

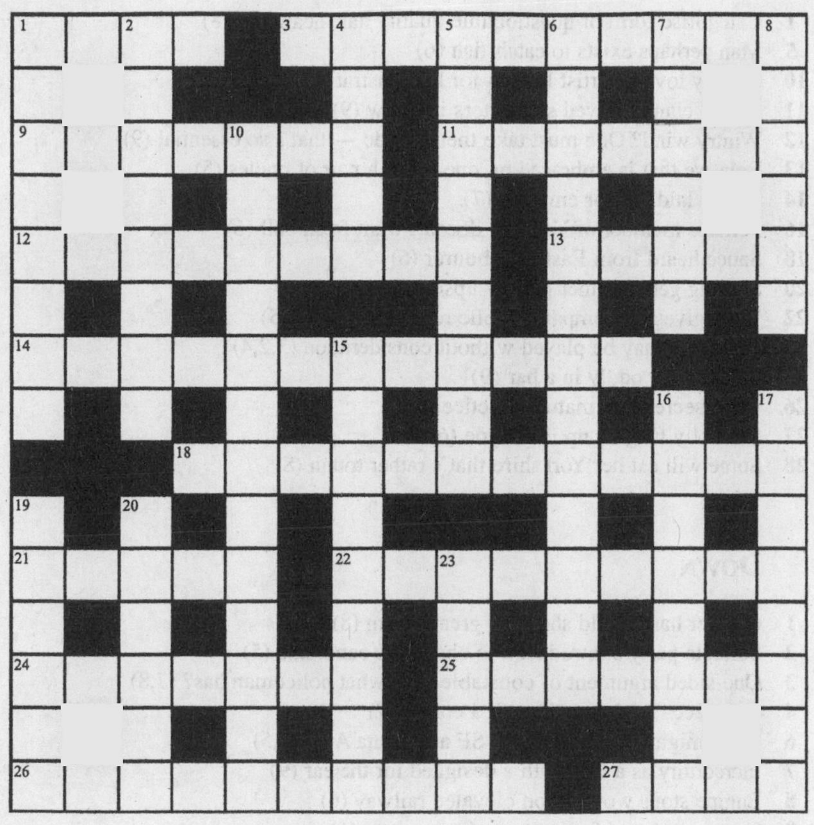

41

Across

1 Categorise form of questionable quality half-heartedly (8)
5 Man perhaps exists to catch fish (6)
10 Greatly love an artist known for his illustrations (5)
11 Doone clan involved supporters in a row (9)
12 Wintry wind? One must take them inside — that's so essential (9)
13 Relative that is embraced by one of each pair of uncles (5)
14 Carpet laid out for empress (7)
16 Perhaps member of Yankees doesn't want fresh ball (6)
18 Sauce heard from Eastern labourer (6)
20 Shilling gets milliner greatly upset (7)
22 What gives oil company public relations answer (5)
23 How card may be played without consideration (3,2,4)
25 Melody lies oddly in a bar (9)
26 Undersecretary's mature practice (5)
27 Sea holly tangled up in groyne (6)
28 Some will eat her Yorkshire that's rather tough (8)

Down

1 Cleaner has a child showing great charm (8)
2 Suitable party's introduced to choose as candidate (5)
3 One-sided argument of constable over what policeman has? (7,8)
4 One needs to change if such a colour (7)
6 What might shift under LA, SF and Santa A? (3,7,5)
7 Incredulity as a labyrinth's designed for the ear (9)
8 Engine stops working on elevated railway (6)
9 Courageous and fortunate under pressure (6)
15 Went with a pony to roam common (3-1-5)
17 Dickens' hero involved in dull and menial work (8)
19 Hill in south-east with unknown level of building (6)
20 One day I dined and got to fill up (7)
21 A pointer to earth growing cold (6)
24 Silver appearing above eroded stone (5)

42

Across

1 Star initially undertakes each month to visit South Africa (9)
6 Second part of poem about artistic figure (5)
9 Completely finished one type of fish, say, in mess (3,4,3,5)
10 Solid surface on which Pachelbel wrote his Canon? (6)
11 Lacking skill? Not in rhyme (8)
13 Deplorable nature of low fees *Sun* adjusted (10)
14 End of movement established by fellow attorney (4)
16 Most unlikely model for a leprechaun (4)
17 Reportedly change composition in church painting (10)
19 It's espoused by people opposed to engagements (8)
20 Constable possibly makes arrest, recalling it for soldiers (6)
23 Free wine seems hollow to express friendly feelings (4,7,4)
24 Compost heap regularly dipped in to transfer seedlings (5)
25 Muse on miners getting right dividend (9)

Down

1 Specialised language used in abusive match? (5)
2 Make renewed effort to shelter inferior members (4,4,5,2)
3 After party in Paris we are very hungry (8)
4 Warrior race not entirely unmatched in Scotland (4)
5 Article compared with continental version of creed (10)
6 Tot's witty saying about university course (6)
7 Sir Thomas wrote it? I'm impressed! It's much better! (5,4,4,2)
8 Speckled bird consumed by picnickers? (9)
12 For example, Ralph Rackstraw's leanings loosely involving the Queen (10)
13 Bat with little hesitation, assuming quiet self-control (9)
15 Plant 20 erected — it's used to transmit power (4-4)
18 Compensate for resting actor's position? (6)
21 Masons' doorkeeper — a revolting person, we hear (5)
22 Pole on film set creating sound of artillery fire (4)

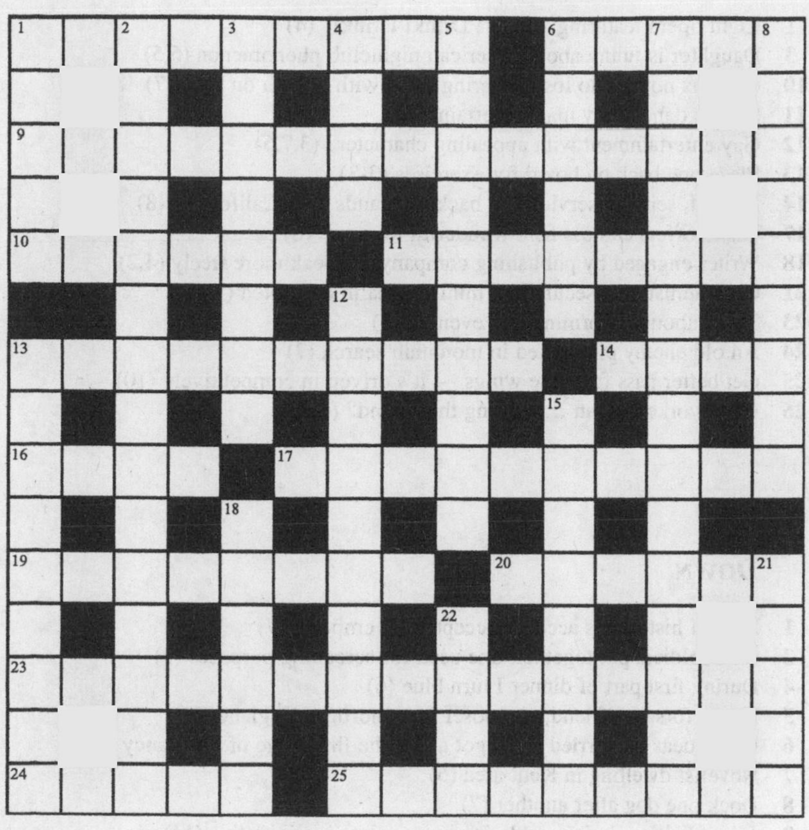

43

Across

1 Light opera featuring "Drink! Drink! Drink!" (4)
3 Daughter is funny about American nightclub phenomenon (5,5)
10 One has nothing to lose, entering mass with wreath on head (7)
11 Square dancing by male entertainer (7)
12 Gay entertainment with appealing character? (3,7,5)
13 Place one back on board for exercises (3-3)
14 Absurd, sending servicemen back to islands near California? (8)
17 Island offers endless fish: wonderful to return (8)
18 Writer engaged by publishing company to speak more freely (4,2)
21 Communist vote secured by minister Stalin corrupted (7-8)
23 Tense about performing this evening (7)
24 An old enemy pinpointed in mountain search (7)
25 Get better pass from the wings — it's driven in competitively (10)
26 Page worker in van … making this sound? (4)

Down

1 Roman historian's account accepted by emperor (7)
2 Plan soldiers put together, one used for screening purposes (9)
4 During first part of dinner I turn blue (6)
5 Henry forsakes friend, composer from north of England (8)
6 False ideas unmarried girl's got about the first stage of pregnancy (14)
7 Novelist dwelling in Kent area (5)
8 Dock one dog after another (7)
9 Join rebellion during work, in manner showing initiative (14)
15 Make arrangement with Conservative personality (9)
16 Richard Murphy said to be a tyrant (8)
17 Mule perhaps located source of marijuana in purest form (7)
19 Pockets gratuity picked up in period of service? (3,4)
20 Wrongly assume he's an Irishman? (6)
22 Vital organ's actual name is concealed (5)

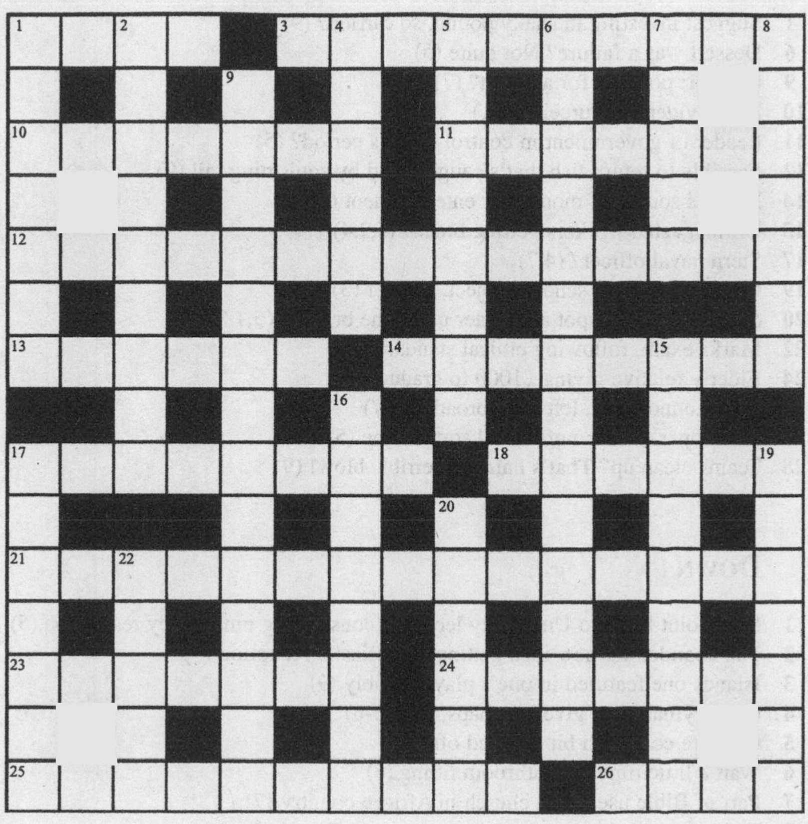

Across

1 Suggest investing in shaky pound, so carried? (9)
6 Dessert was a failure? Not quite (5)
9 Cleaner: position for an idiot? (7)
10 Note evident in Purcell? (6,1)
11 Leader of government in control for this period? (5)
12 Feasible to return fish that's caught, held by squirming tail (9)
14 Endless source of money for entertainment (3)
15 US magazine workers getting bread? (5,2,4)
17 Stern naval officer? (4,7)
19 Covering 60% of school subject, in brief (3)
20 Suspect one can spot a deserter under the bridge? (5,1,3)
22 Marks exam, following ethical standard (5)
24 Elderly relative giving £1000 to graduate (7)
26 John Lennon, say, left after broadcast (7)
27 From opera house unfinished sort of tune (5)
28 Teams clean up? That's hardly a terrible blow! (9)

Down

1 Sore point leads to University lecturers considering emergency resolution (5)
2 Pub abandoning new idea getting enthusiastic reception (7)
3 Islands one featured in one's play, possibly (9)
4 Countryman may give her maps, free (5-6)
5 Obscure coin with bit chipped off (3)
6 Wait a little time for bathroom fitting (5)
7 Part of Bible used by a church in African country (7)
8 Former criminal group sent for execution? Great! (9)
13 Starting bridge tournament from personal experience (2,5,4)
14 It allows you to predict how gift horse may turn out (9)
16 100% sated with what one hears from Communist? (4,5)
18 Assert historical period to be just like any other (7)
19 Italian scientist to spoil study on electric current (7)
21 Fix coursing at last through vein (5)
23 Happy to miss first place, showing great athletic ability (5)
25 Girl offering no opening for oaf (3)

44

45

Across

1 Suggestion of whip in party game? (6,2,6)
9 Top university job filled by combination of results (9)
10 Clergyman curtailed sidc (5)
11 Thin character hugged by fat, rich Scotsman (5)
12 As legend would have it, models lick floor (9)
13 Most mean to remove opener from contest in cricket match (8)
15 The spirit of Nebuchadnezzar? (6)
17 Male with weak back needs injection of boron to preserve body (6)
19 An orderly order? (5,3)
22 Thousand trapped by ludicrous price in trouble, subject to process (9)
23 Yank no good after exile in African languages (5)
24 Return of sailor I love figures in relationship (5)
25 Tragic design, and disastrous (9)
26 Destination for victims of cabinet reshuffle? (6,8)

Down

1 It could be rude! (4-6,4)
2 Criminal framing wife in fix gets bird (7)
3 As some boats flew, seconds lost (5)
4 American swimmer's low fins switching height (8)
5 In recent times, rent raised to cut deposit (6)
6 One done in spring (5,4)
7 Literary taste but about right (7)
8 Once mistaken about island, personality theorist left area with bleak outlook (8,6)
14 Lens takes little animal from big, *in toto* (9)
16 Leaves having no function inside trunk? (8)
18 Bishop requiring one, milliner's heading off, turning up with this? (7)
20 Payment offering little money — is none unexpected? (7)
21 Quite a serious problem's contained (6)
23 Maybe we hear palm trees are alongside this tree (5)

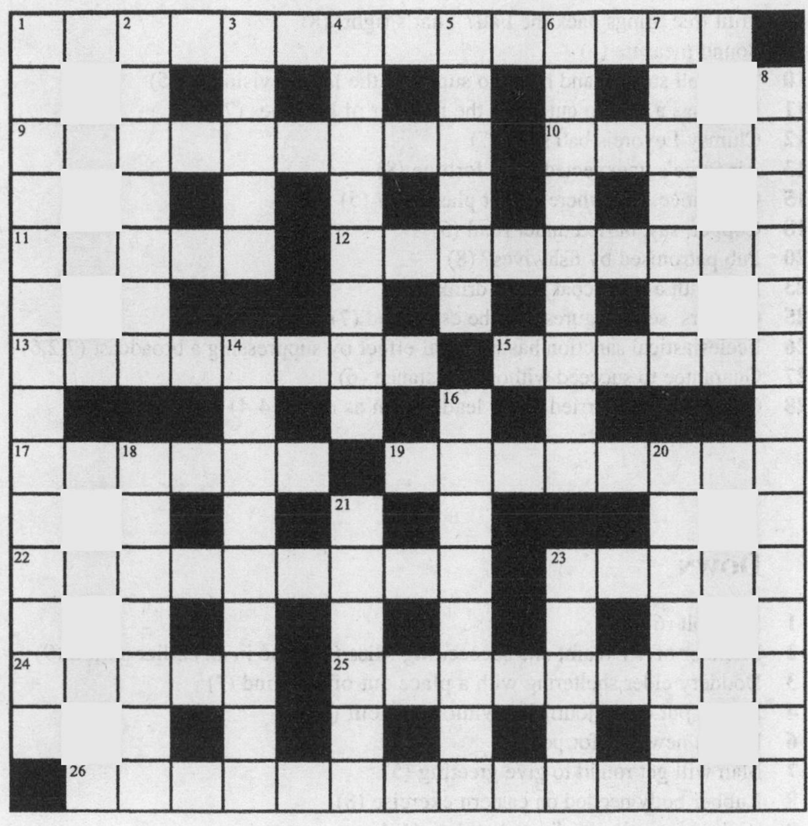

46

ACROSS

1 Fruit tree brings back the Fall? That's right! (8)
5 Sound measure (6)
10 They tell stories, and hurry to suppress the least revision (10,5)
11 Head has a way to cut down the number of branches (7)
12 Clumsy Eeyore's bad sight (7)
13 Air force's unexpected good fortune (8)
15 Old prince, and where he got plastered? (5)
18 Copper, say, buried under road (5)
20 Pub patronised by fishwives? (8)
23 Dog with a short coat needs drink (7)
25 Gunners' set of figures may be estimated (7)
26 Ecclesiastical sanction has regional effect by suppressing a broadcast (7,2,6)
27 Guarantee to succeed without resistance (6)
28 On reflection, married youth leader seen as naive (4-4)

DOWN

1 Slip bolt (6)
2 Member of TV team, one succeeding Miss Gwynne in an earlier period (9)
3 Doddery elder sheltering with a place out of the wind (7)
4 Firmly put down journalist without honour (5)
6 Design new part for port (7)
7 Man will get round to give greeting (5)
8 Rubber body needed on eastern exercise (8)
9 Right winger is a dedicated athlete (4-4)
14 A recent vehicle secured, on special order (1,2,5)
16 File to cover up orally with a sign of disapproval (9)
17 Muscovite taken in by expert showing friendliness (8)
19 Star striker (7)
21 Act better in old-fashioned drama (7)
22 Changed colour: it faded again, we're told (6)
24 Mysterious letters, some for UNESCO (5)
25 Loot left in epidemic (5)

46

47

Across

1 He looks at lots of characters, to interpret one (12)
9 Fish — that is to say, perch? (5)
10 A second chap's drowning in the drink, an arm of the Pacific (6,3)
11 No firm believer should be acting so badly (8)
12 A bit taken off wood as discount (6)
13 In 1940, Norway was busy (8)
15 Cruel person without right seat of emotions (6)
17 Attribute shown as I am given place at Cambridge finally (6)
18 One that could be flaming or left in depression meeting a politician (8)
20 Fate makes idiot call for attention (6)
21 Like holy tirade expressing high ambition? (8)
24 Rage when in drink, here's the essence (9)
25 Allow the man to show forgetfulness (5)
26 Female — one told story about guy and played for time (12)

Down

1 Police try to impound doctored tapes (7)
2 Big paranoid cur barking in a domestic closet (6,8)
3 A place of some darkness, first to last! (5)
4 Castle — it is arranged in criss-cross patterns (8)
5 My turn — quiet please! (4)
6 Son to worry about rebuke — he's shabbily dressed (9)
7 Troubled, as one responsible for tiles having a tile loose? (2,3,2,1,6)
8 Old man needs shelter — that's obvious (6)
14 Up at loch, curious to find one scented shrub (9)
16 Benediction following an unexpected expulsion (5,3)
17 Ancient graduate in church taking part (6)
19 Place primarily wanted by the man stuck in adit, struggling? (7)
22 Bay home already taken? (5)
23 African money kept in rice dish (4)

47

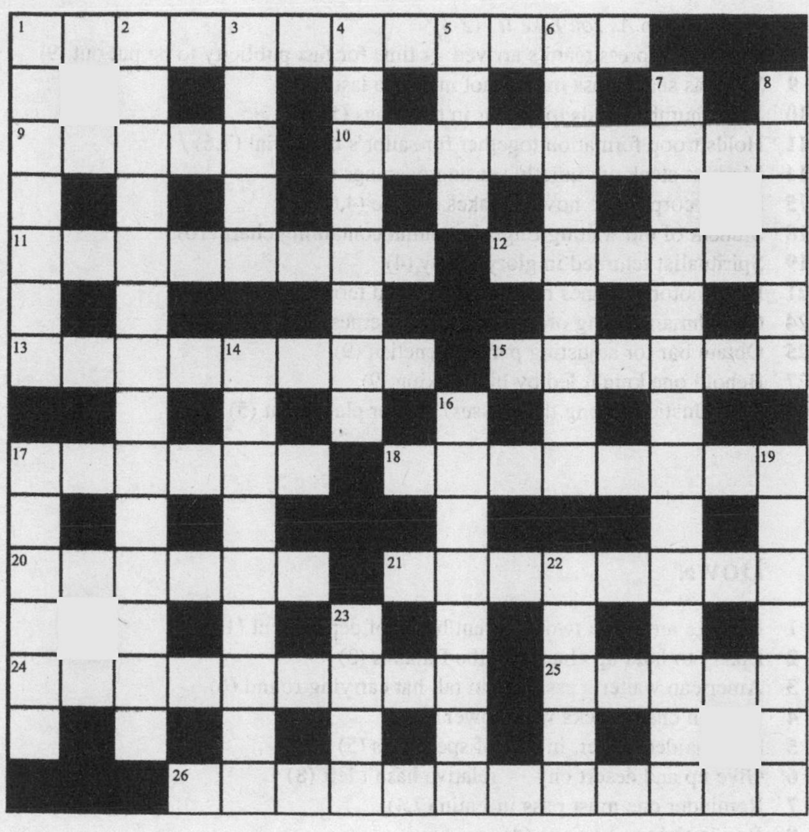

Across

1 Improvise in *As You Like It*? (2-3)
4 Member of press team's arrived — time for fast publicity to be put out (9)
9 Work as seamstress maybe not meant to last (9)
10 Some number leads to doctor in two rings (5)
11 Holds troop formation together for sailor's memorial (7,6)
14 Markets stock up, quietly sensing openings (4)
15 Flying corps hope novice makes college (4,6)
18 Student of old writing English column containing chart (10)
19 Spiritualist returned in glory, oddly (4)
21 Here motorist comes by train, plane and ferry journey (5,8)
24 Churchman joining order generates bitterness (5)
25 Obtain bar for adjusting part of Venetian (9)
27 Behold one knight led by bird to king (9)
28 Chief Justice among the masses? A poor placement (5)

Down

1 Manage area with two different heads of department (10)
2 Likely to hold up change in the Balkans (3)
3 American waiter's assistant in tall hat carrying round (6)
4 Band in charts rocks with power (9)
5 Saw maiden, sober, in pair of spectacles (5)
6 Give up and desert city — relative hasn't left (8)
7 Reminder one must pass in Latin (7,4)
8 Boy is right to keep up (4)
12 Glasses only needed for this daytime course? (6,5)
13 Knowing good fortune comes about through deadly craft (6,4)
16 Determine early broadcast aired porn (9)
17 With clipped accent fabricate mean accusation (8)
20 So friendly with crowds everywhere (6)
22 Sound from cheap pub relating to earpiece? (5)
23 Form of office in Washington (4)
26 Old Indian crown cheers leader in Jaipur (3)

48

49

Across

1 Response with matching offer (15)
9 As hands sweep round, wick crumbles in the end (9)
10 Chaps changing sides in state capital (5)
11 Dim baronet falls into river (6)
12 Silicone Valley? (8)
13 American in English school is 21 (6)
15 Calf-length? That's rather short (4-4)
18 Factory engaged in bottling large number of marine organisms (8)
19 Mouse, a device that lightens one's work (6)
21 *Depot* is the American expression (8)
23 Fast-paced thriller writer? Yes and no (6)
26 Sharp sound from wife bitten by circling insect (5)
27 Christopher, one employed by Nottingham Forest (9)
28 Pseudery shown by religious head ringing about wine (15)

Down

1 Series of events surrounding active hurricane (7)
2 Marauder's about to be sunk (1-4)
3 Make an appraisal, as shoplifters do (4,5)
4 Check only one kidney? (4)
5 Tall grass and heather sufficient for bird (8)
6 Dish concocted by Dad's Army (5)
7 South Africa installing large tank in animal sanctuary (9)
8 Find one innocent in connection with the murdered Archbishop (7)
14 Leaders in space technology expose incident in flight (9)
16 Stars' table has oriental fairytale beauties round edge (9)
17 Pandering to public taste, fizzy drink's on university menu (8)
18 Trophies end up with fastest team here (3,4)
20 Ornate rose-red screen (7)
22 Top missing from unique fireplace (5)
24 English support for toilets being free (5)
25 Drive off when filming's finished early (4)

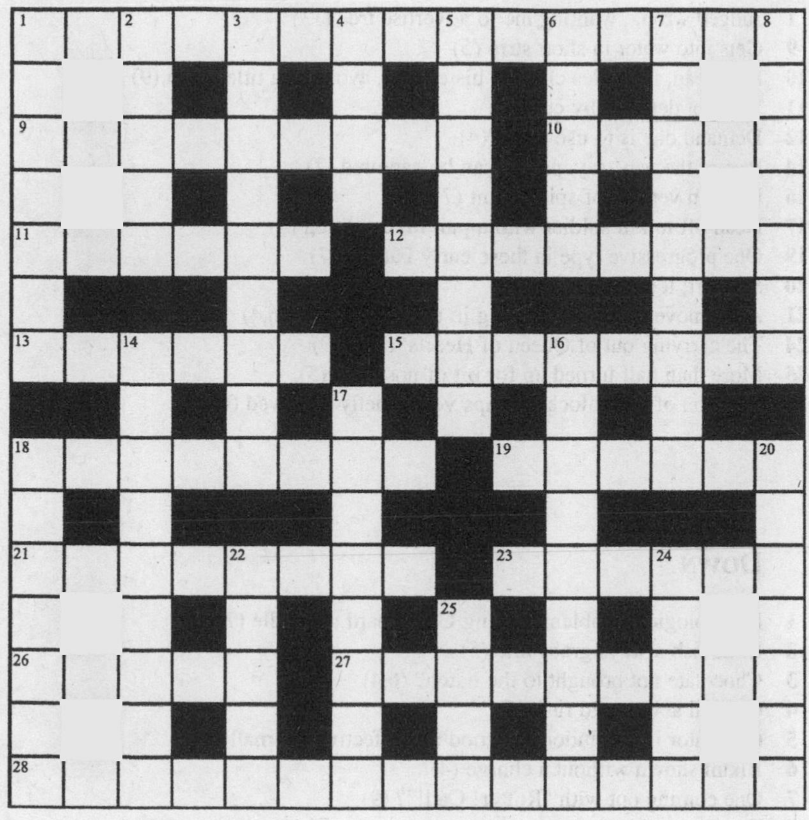

50

Across

1 Judged wrong, wanting me to advertise free (13)
9 Gets into water in short strip (5)
10 European, in circles close to his capital, avoiding a titled man (9)
11 Was not deterred by cut (10)
12 Demand one is to use apron (4)
14 Part of the way, say, pawns can be captured (7)
16 Utopian version of spiny plant (7)
17 Head off to kill soldier with a painful condition (7)
19 One progressive type in these early Fathers (7)
20 Support; it sounds equal (4)
21 After move, adult fits nothing in this tiny space? (6,4)
24 The carrying out of Queen of Hearts' order (9)
25 More than half turned up for bit of nostalgia (5)
26 Creation of road block perhaps very quietly reviewed (6,7)

Down

1 Psychological problem making Greek hard to handle (7,7)
2 Little fish always grabs line (5)
3 Chocolate not brought to the hatch? (6,4)
4 Council about gold rush (7)
5 Generator is revolutionary, good and effective internally (7)
6 Bikini shown without a charge (4)
7 One coming out with "Rotter! Cad!"? (9)
8 The French king wants a corps to have powerful trumpeter (5,9)
13 Speculator investing nothing in the more exotic? Give it a whirl! (10)
15 Kentish Town cemeteries full? (9)
18 Nest piled high, left new 22 in a mess (7)
19 How pompous Dickensian got at the peas? (7)
22 Piece of wood I found to be police baton (5)
23 South African unknown to have been on the radio (4)

50

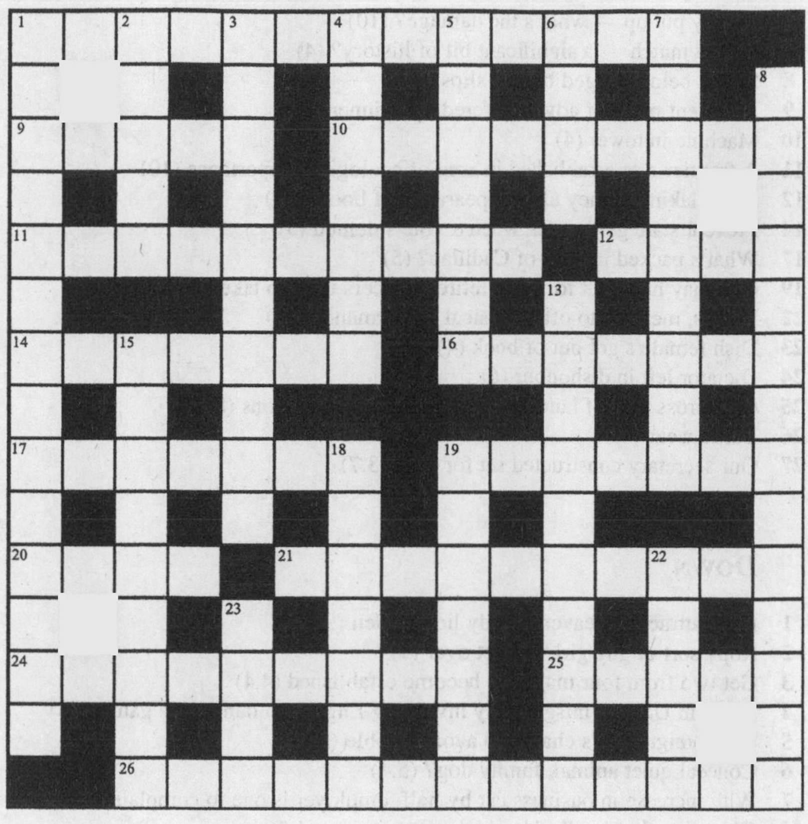

51

Across

1 Money put up — what's the damage? (10)
6 Boy at match — a significant bit of history? (4)
8 Escort being hugged by girl slips (8)
9 Excellent piece of advice offered by spinner (3-3)
10 Machine in tower (4)
11 A frontier's re-established in area of ecological importance (10)
12 Poet making money after appearance of book (4,5)
14 Prevent state getting tax when a vote's denied (5)
17 What's packed in boot of Cadillac? (5)
19 One may have felt for what retired officers used to take on (6,3)
22 Lyrists, men out to offer musical performance (10)
23 Dish female's got out of book (4)
24 Dictator left in dishonour (6)
25 Cut across area of land having internalised directions (8)
26 Battle wear (4)
27 Our secretary constructed set for play (3,7)

Down

1 Most attractive heavenly body lies hidden (9)
2 Ropy sort of guy girl will get over (7)
3 Get two from four maybe to become established (4,4)
4 Restrain US lout misguidedly involving English in dangerous game (7,8)
5 The foreign gent's chance to avoid trouble (3-3)
6 Conceal quiet animal, family dog? (5,4)
7 With increase in business cut by half, employer is one to complain (7)
13 Older female friend's friend is an easy target (4,5)
15 Report of 19 shoddy items? Retaliation called for (3,3,3)
16 Put floor covering down in gym for friend (8)
18 When in trouble I resort to bluster (7)
20 Nellie is embraced by explosive old Greek (7)
21 Mark time as German politician (6)

51

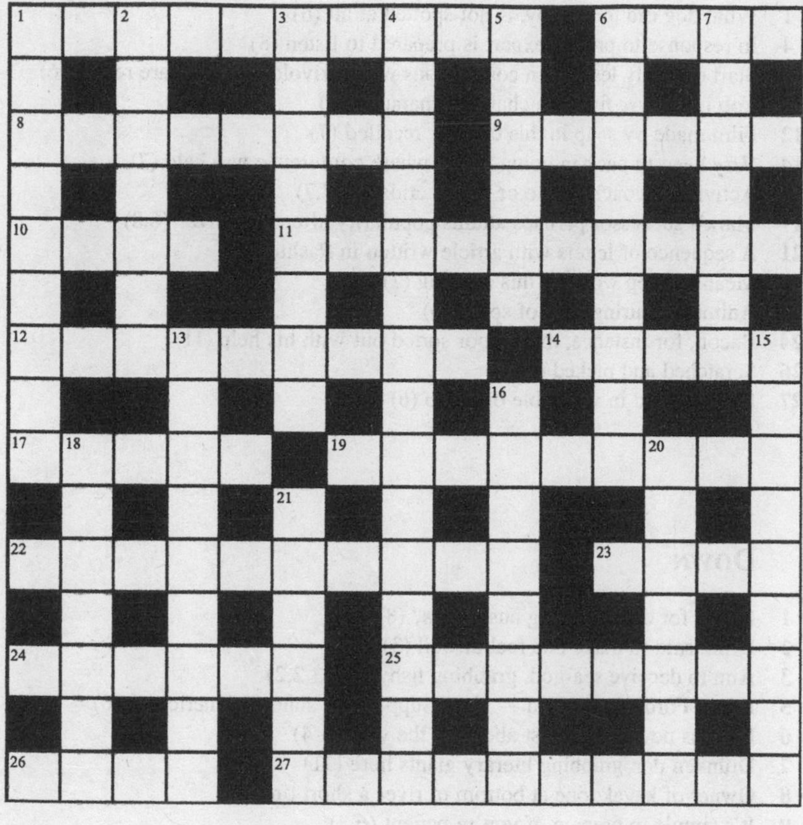

Across

1 What dog did to cat, say, is not spotted at all (6)
4 In response to prayer, expert is prepared to listen (8)
10 Start off daily lessons in convulsions when frivolous reports are read (5,6)
11 Yob in *Grease* finally a changed character (3)
12 Film made by soap in this city bar recalled (7)
14 Very keen to pack in going round where conference was held (7)
15 Activity of couch potato or booze cruiser? (7,7)
17 Mark's successor perhaps attains popularity after forty-five? (6,8)
21 A sequence of letters with article written in Pashto (7)
22 Mean to keep working this evening (7)
23 Animal requiring a lot of space (3)
24 Bacon, for instance, that's poor sorted out with his help (11)
26 Scratched and picked (8)
27 Decay found in vegetable or shrub (6)

Down

1 Ready for card-carrying customers? (8)
2 Drink said to make one feel unwell (3)
3 Aim to deceive sea-god, grabbing fishy tail (3,2,2)
5 Henry Ford's brainchild — show support for state of America (4,10)
6 Endless new work's just about on the way (3-4)
7 Drunken don grabbing literary giants here (11)
8 Owner of kayak, one at bottom of river a short time (6)
9 It's simple to prepare, if you're patient (6,8)
13 Grand spot? It might be, for general store (7,4)
16 Panic originally associated say with their letters (8)
18 Got the picture from file *Guardian's* foremost journalist kept (7)
19 Bitter resentment that's expressed in private (7)
20 Be in dispute with bank in town at start of March (6)
25 The shot to beat — a winner (3)

53

Across

1 Person of authority is hard, but quite fair (8)
5 It's a bullet, you idiot (6)
8 Unitc to make cabbies go out of business? (5,5)
9 Scotsman's brief word of thanks spurned by charity organisation (4)
10 Digitally-processed picture? (6-8)
11 Serviceman repulsed fighters with gun (7)
13 Badly-behaved thug, after treatment, is in denial (7)
15 Pull vehicle backwards through a series of junctions (7)
18 Italian town in country area accommodating soldier (7)
21 Some Americans condemned nation's dirty habits upfront (14)
22 South African fighter losing heart (4)
23 Organise dodgy trade importing cars (10)
24 Saw British expedition leader taken captive (6)
25 Reduced to penury, flighty creature has retired? (8)

Down

1 Devil flares up when struck (7)
2 Virtuous gentleman announced retirement wish (9)
3 Fellow with book coming in to recount Iris Murdoch story (3,4)
4 Escape in this car, perhaps (7)
5 Head of department is taking English period, being free (9)
6 Non-U greeting in early part of day (7)
7 Joanna's support (7)
12 Companion interrupting a rude fellow (9)
14 Said hello to convert in part of church (4,5)
16 Power to twist cable (3,4)
17 Player about to hit ball (7)
18 Saint's cherished religious teachings one initially ignored (7)
19 Having phoned home, start to go travelling far and wide (7)
20 Confident lower classes will join country uprising, I left (7)

53

54

Across

1 Realizing endless oil could make chap stupendously wealthy (11)
7 Nothing odd in owning a rug (3)
9 Nobody makes escape after defeat (9)
10 Is it broadcast as billed? (3,2)
11 Slight rolling of r's not finished with by way of introduction (7)
12 Scenic picture from satellite brought into play (7)
13 Church affirmative about decline, on reflection (5)
15 Slow in terrible heat, coming in weakened (3-6)
17 City's retrograde step, surrounding line judge initially (9)
19 Parts of the kingdom need some prophylactics (5)
20 Flatten and clean fish ready for eating (7)
22 Touching rendition of *Cat in the Hat*? (7)
24 Climbers struggle in extremes of iciness (5)
25 Novice brought in by Middlesbrough banker to manage contract (9)
27 Leg in contest European lost (3)
28 There's more than one to a cell in convent (1,10)

Down

1 Vitality's answer for one making kill (3)
2 Artist using brass for ears (5)
3 Fashionable paper folded before finishing year's investigation (7)
4 Some thread is necessary for binding foreign article (9)
5 In Association missing Liberal is given black mark perhaps (5)
6 Set off to walk slowly into river (7)
7 Identical paths, one taking short route left, perversely (9)
8 Compromises vegan diet, mistakenly eating a small piece of kidney (4-3-4)
11 Idiot rushing around making cry (4-7)
14 Metal handle for girl one put in chimney (9)
16 A test for vehicle climbing up to old Mexican plant (9)
18 Lesbian seems confused about a kiss (4-3)
19 Put together illustration like true artist (7)
21 Cape in essence a waterproof coat (5)
23 Fine introduced in club for chopping tree (5)
26 4 may pass through this study (3)

54

55

Across

1 Less conventional film director (6)
4 Wear shoe with flimsy lining (8)
10 High on drugs, son walked away from home (6,3)
11 Crazy, say, about a girl (5)
12 Scottish team's top player given red card? (3,2,6)
14 Cut-price holiday destination (3)
15 Swiss town near river is not entirely secure (7)
17 Left with one pound in cash (6)
19 In some ways, love can produce a feeling of pity (6)
21 Early parts of course deferred for apprentice? (7)
23 Odd characters employed by Afghan commander (3)
24 Blue gear guy sported for game (5,6)
26 African having little time in part of London (5)
27 Most dependable American boat finally enters port out east (9)
29 A high explosive used in raids blowing up military bases (8)
30 Festive occasion attended by second pair of sexy stars (6)

Down

1 Wife pitches into drinking bouts (8)
2 Seal changes to European contract (5)
3 Poor wee creature (3)
5 Shoot dead gangster after resistance (7)
6 NY location in which this crossword grid appears? (5,6)
7 Fiscal period, in a manner of speaking (9)
8 Original Olympic setting as discussed for film (6)
9 Fruit yours truly provided halfway through game (6)
13 Storm engulfs the Royal Marines as well (11)
16 Individual qualities a Conservative put into constitution (9)
18 Very little money in bank of late (8)
20 Given a place, say, with a view? (7)
21 Sample fish stuffed with something unknown (3,3)
22 Tree, source of spice in cold continent (6)
25 Woman implicated in Watergate upset (5)
28 Schoolmaster less than halfway through meal (3)

55

56

Across

1 Someone always ready for action — General Tom Thumb, perhaps (9)
6 Naval officer gets Italian to come clean (5)
9 Lizard for instance repelled by cold floor (5)
10 A pet prone to wander — it might carry a message (9)
11 Bell showing time to tuck into pudding — British one (9,6)
13 Difficult to understand returning to gallery, but good luck (8)
14 Accommodating one in factory (6)
16 You and I encounter black beetle (6)
18 Where you might find pal I shot, injured (8)
21 Clear choice for one tempted to thieve — no more concessions (4,2,2,5,2)
23 In deep water? Not if you keep your footing (2,4,3)
25 Last to use unpleasant resin (5)
26 Drink everything? Not quite everything (5)
27 See our way diverted back to square one (2,3,4)

Down

1 Great coat worn by soldier (5)
2 Bill and coo, side by side (4,3,4)
3 Author said to be a person of dubious morality (7)
4 Nanki-Poo, for instance, established church out east on line (8)
5 Jewel container turned up in tree (6)
6 Clothing to provoke outrage when worn by relative outsiders (7)
7 Clean hair (3)
8 Plant used to cure people suffering endless harm (9)
12 In the Newcastle area, clay's black and blue (4,3,4)
13 Swaps — two I ought to pass up (2,7)
15 Dog on the line — it's mine (8)
17 Within it I always include a shortened version of my name (7)
19 Table recently found in French city (7)
20 Turn singer into a bareback rider (6)
22 Plant pronouns together (5)
24 Toddler starts to trip over things (3)

56

57

Across

1 Direct from prison enclosure, in trim beyond compare (3,10)
9 Ohio city area built with Swedish money mostly? (5)
10 Fake burn, Hollywood's skin colouring? (9)
11 Extraordinarily communicative primate allowed to turn almost hoarse (10)
12 OK, I must abandon funniest attempt at humour (4)
14 Answer concerning South Atlantic ... (7)
16 ... must have upset querier (7)
17 Female with the most casual manner cut good people (7)
19 Distraught father, old-fashionedly ludicrous (7)
20 Pretentious salver, almost entirely overturned? (4)
21 Healthy congratulations (4,3,3)
24 Unreasonable Illinois denizen imprisons soldier (9)
25 The object is to keep backing writer that lacks skill (5)
26 Fair play abandoned, so old cunning accepted, at this time? (5,5,3)

Down

1 Mould for arts applies ... this? (7,2,5)
2 Bucolic atmosphere not initially penetrating river and lake (5)
3 Chip ox bone apart, hating foreign bodies? (10)
4 In a stir, being previously summoned to court? (7)
5 Pile a fiery allotment with more foliage (7)
6 Legally ineffective letter in Greek will (4)
7 In Italy, three million are outwardly fanatical (9)
8 Novel on the contrary about upper class is one of Baldwin's (7,7)
13 European, as against Iranian money, like a Guinea? (10)
15 Caught in the wrong, work a little bit (9)
18 Resigned from time in party after central reshuffle (7)
19 Opera running smoothly, even though held up? (7)
22 Amount produced from regular samples of pyrites lode (5)
23 Work involved in *Rite's* half over, making Stravinsky's name (4)

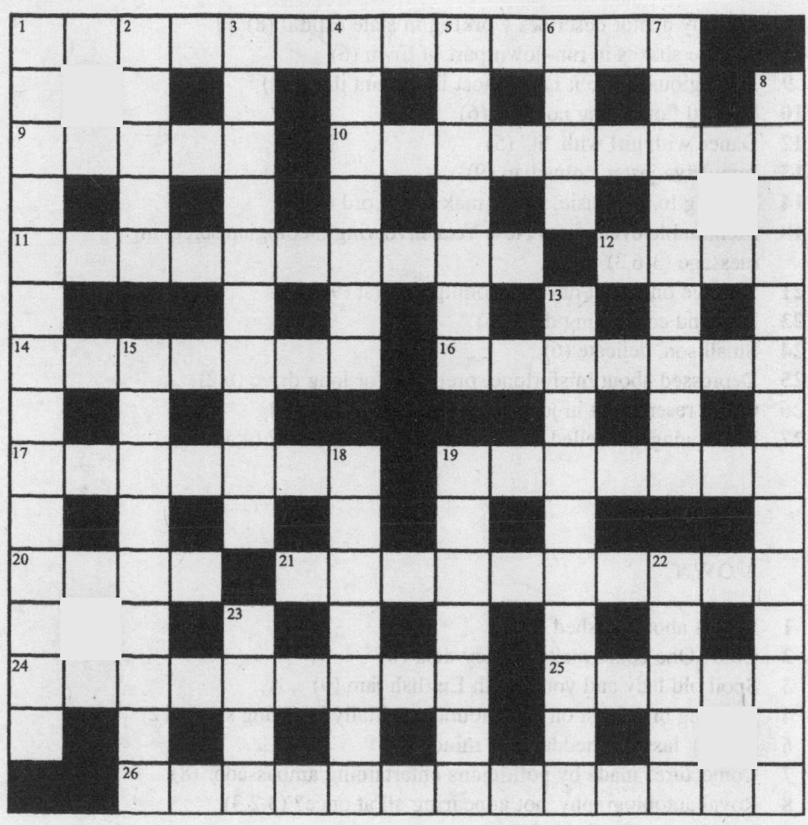

58

Across

1 Wealthy doctor describes working in state capital (8)
5 Got the shakes in run-down part of town (6)
9 In religious house it is the most important thing (8)
10 Recited flourishing novelist (6)
12 Dance with girl with "it" (5)
13 Impulsive jester, comedian (9)
14 Moving to the music, group making record (7,5)
18 Memorable event from New Year involving theologian screening message (3-6,3)
21 Sausage one must have seasoning on first (9)
23 Respond concerning deed (5)
24 Small son, delicate (6)
25 Depressed about misfortune, prepared for long drive (6,2)
26 Cause resentment in joint after stripper's finale (6)
27 Right-winger levelled — what do we hear ref did? (4,4)

Down

1 Meal's about finished (6)
2 Cold? One could make a spicy dish (6)
3 Spoil old lady and youth with English jam (9)
4 Warning broadcast on it, announcer initially breaking story (12)
6 Grating last of Cheddar into mince (5)
7 Conjectures made by politicians entertaining ambassador (8)
8 Royal autobiography, not appearing all at once? (3,2,3)
11 Washington Heights? (5,7)
15 Trouble during first half of reel after square dance (9)
16 One making toast with reference to a departing Queen (8)
17 A party overcomes tricky point in formal selection of candidate? (8)
19 Writing to the Spanish composer (6)
20 Short journey announced that crosses Siberia (6)
22 Woman not wholly above the law? (5)

58

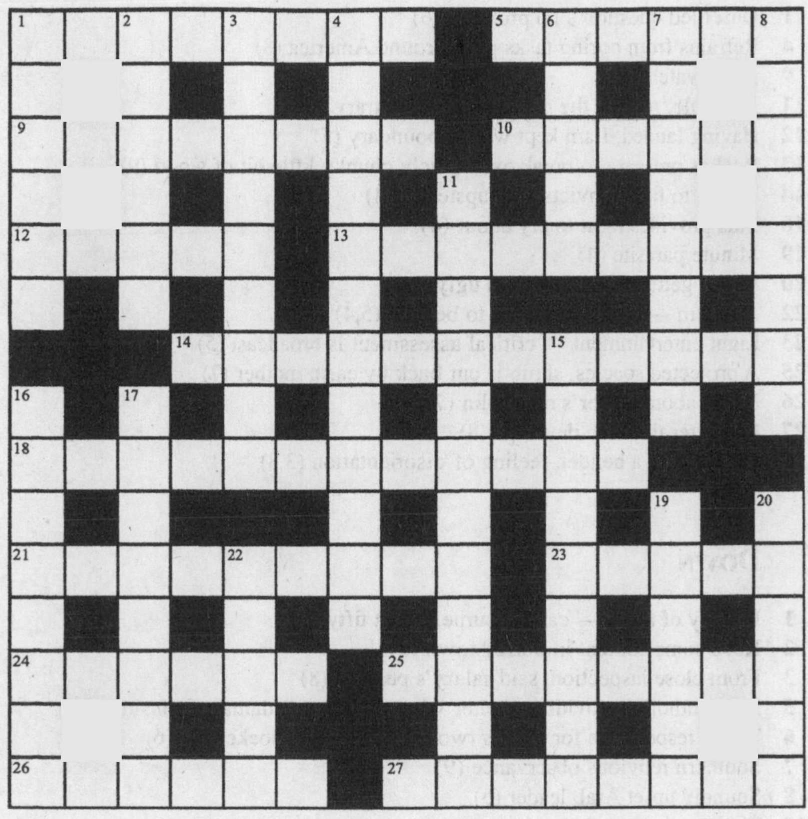

59

ACROSS

1 Unsettled question's no problem (6)
4 Refrains from boring tasks going round America (8)
9 Keep watch (7)
11 As a rule, rake in the dosh? On the contrary (7)
12 Having landed, I am kept within boundary (5)
13 It takes pressure to break excessively chunky little bit of wood (9)
14 Where to find convicts and tapsters (6,4)
16 This provides light to cry about (4)
19 Minute parasite (4)
20 Writer getting older — plain ugly (10)
22 A truism — there's no more to be said (5,4)
23 Light entertainment — critical assessment is broadcast (5)
25 A protected species, shrub is cut back by earth mother (7)
26 Novel about leader's old violin (7)
27 Hereafter tiny tree develops (8)
28 Let loose in a bender, feeling of disorientation (3,3)

DOWN

1 Medley of tunes — can I assume, about fifty? (9)
2 Keep some content in Surrey town (5)
3 From close inspection, said salary's peanuts (8)
5 Port Authority providing shelter when her mast is damaged outside (7,6)
6 Thugs responsible for town's two leaders being knocked off (6)
7 Southern religious observance (9)
8 Soundly upset Arab leader (5)
10 The man is blest, being incorporated in this (13)
15 Horse given a further bridle (9)
17 To show disapproval of small cake, said "It keeps coming back" (9)
18 Competent girl comes in early, winning (8)
21 Like a seabird, following behind vessel (6)
22 Toledo's first sword found in a tent (5)
24 Outspoken, very often coming across like leaders (5)

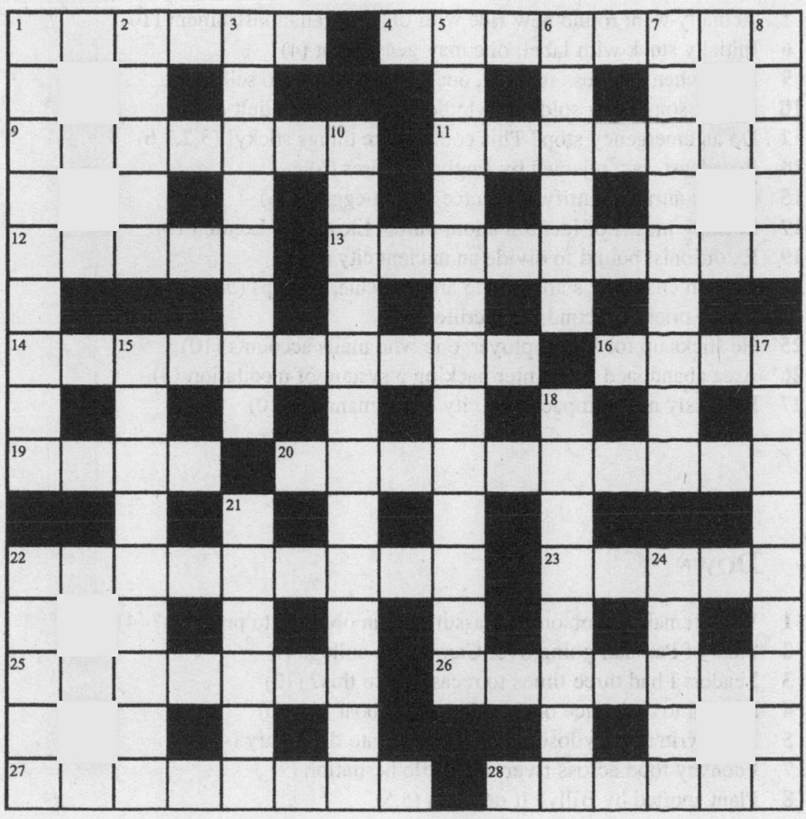

60

Across

1 Actually went round new ride with old Australian instrument (10)
6 Initially stuck with label, one may get in a rut (4)
9 Even when business is slack, one may have a lot to sell (10)
10 Return something sold by swindler that's not yet adult (4)
12 Do an emergency stop? This could make things sticky! (3,2,3,6)
14 Broadcast, say, rejected by English adviser (6)
15 Main contract identifying source of fish-eggs? (3,5)
17 Lacking means of identification, unlike Lloyd's of London (8)
19 Extortionist bound to divide an ancient city (6)
22 Peevish character's allusion to another clue, perhaps (5-9)
24 Lamb, priest's second for sacrifice (4)
25 He sticks up for his employer, one who mails accounts (10)
26 Area abandoned by painter backing a system of meditation (4)
27 Endlessly make impact on a city opera manager (10)

Down

1 What female impersonators assume is an obstacle to progress? (4)
2 State of Parisian going over Cambridge college (7)
3 Leaders I had three times to recast, to do this? (12)
4 Natural to announce one's selected for boat race (6)
5 Alert writer easily loses heart dipping into dictionary (4-4)
7 I convey food across river with little hesitation (7)
8 Plant sported by Billy? It depends (5,5)
11 Absurd notice displayed outside theatre by old American (12)
13 Over-precise lookout unknown outside European navy (10)
16 It draws attention to passage some master is marking (8)
18 Low circular enclosure where boat is secured (7)
20 One may undertake delivery of a variety of lines (7)
21 Creditor seizing Fitzgerald's stock of wine (6)
23 Ancient ship's lading? Not at first (4)

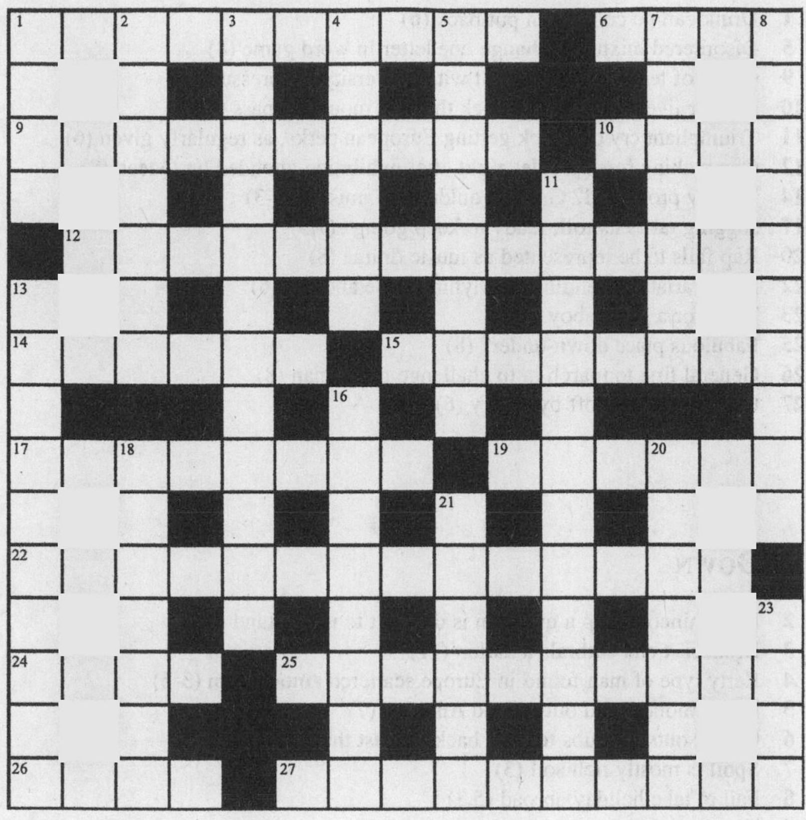

61

Across

1 Drink can go cold when put back (6)
5 Disordered mixture? Change one letter in word game (8)
9 Chary of tending to find fault with university for pressure (8)
10 Pen for livestock arriving back through mountain pass (6)
11 Triumphant cry of Greek getting European perks, as regularly given (6)
12 One making foreign sales right after exhibition attended by Queen (8)
14 Unduly prolonged? Groan wouldn't go amiss (4-5-3)
17 Jogging takes its toll; ready to keep going? (7,5)
20 Rap fails to be represented as music drama (8)
22 Bliss variation's ending on rhythmic rise and fall (6)
23 Term for a girl or boy (6)
25 Fabulous place down-under? (8)
26 General first to march in to challenge policeman (8)
27 Carriage headed off by Henry (6)

Down

2 From nincompoop a question is difficult to understand (6)
3 Broadcast end enthrals a nation (11)
4 Early type of man found in Europe scattered among corn (3-6)
5 Delay money paid out around America (7)
6 Gather outside clubs to look back on past things (5)
7 Spoil is mostly rich soil (3)
8 Fail to take holiday abroad (5,3)
13 Wagner epic has a harsh critic — though nuts are totally engaged with it (4,7)
15 Saucy line he's put into play? (9)
16 Medicine needs to be prompt that's swallowed right at four (8)
18 Artwork featuring headless parrot inside wire bars (7)
19 English arriving in very pleasant Adriatic city (6)
21 Entrance hall for bringing in the old (5)
24 Answer I would support (3)

62

Across

1 Case for diamonds, perhaps (4)
3 Example of words initially exchanged in New College, Oxford? (10)
10 Tree old American sacrificed in a harsh song (9)
11 Compère making gong sound (5)
12 In Bow, chaos at a party — one may be stoned (7)
13 Refuse to do without British rock music (6)
15 Muscle used by sultanate represented within European novel (3,5,2,5)
18 In which there's current support for human (and canine) progress (5,3,2,1,4)
21 Black convict's fatigue after flight (3,3)
23 Terrible article from pen of Sergeant-Major Jolson (7)
26 Compound always divided by public way (5)
27 English PM accepting honour is framer of abstracts (9)
28 Firm in Paris, one producing … (10)
29 … clothing ultimately usable in conflict (4)

Down

1 Hanger-on left first part of play in tizzy (10)
2 Musical princess with house in silver-mining state (5)
4 Attendant admitting crime in cleric's place (9)
5 Primate in royal house out East (5)
6 We may not approve when this arch is raised (7)
7 Single soldier finally going into manoeuvres in disguise (9)
8 Come up to counter (4)
9 Book of the month presented to Bible circle (6)
14 Fan requiring apparatus to assist breathing (10)
16 Primitive retreat entered with serious intent (9)
17 Part of Thomas à Kempis's work produced by copyist? (9)
19 Blow up general in manoeuvres (7)
20 Upset drink, going round via old Egyptian town (6)
22 Continental king put out by duke's covetousness (5)
24 Oil producer, one unknown in West (5)
25 Live with a socially acceptable boyfriend (4)

63

Across

1 Three rivers, one discharging copiously (6)
5 Harry Lime's position in the field (5,3)
9 The keeper is an old servant (8)
10 Showman almost stupefied after ban (6)
11 One may be burning to cheat admirer (8)
12 Motorway's access is not a suitable one (6)
13 Farewell to the double feature (4-4)
15 Move stealthily, dumping swag at the front, then run slowly (4)
17 Force a boundary (4)
19 How bad the advertiser in the personal column used to feel? (8)
20 Gunners face moving from the centre (6)
21 Expose new grandee as crook (8)
22 Language applicable when English not available (6)
23 Jat reverts to halma-playing in the mausoleum (3,5)
24 One trying to influence jury to accept ruffians' leader (8)
25 Officer cannot hope for rise to such a rank (6)

Down

2 Put on too much weight to organise a do with lover (8)
3 Tail-end computer devices for night fliers (8)
4 Putting in for payment for plaster (9)
5 Address to audience by proselytiser produces transfer of power (6,9)
6 Angler with one alternative: to take a new line (7)
7 Chamber for the maths set (8)
8 It shows who should be wearing the trousers, for example (4-4)
14 Sort of crime apparent during border pursuit (6,3)
15 Connive in liaison (8)
16 Crustacean may be barnacled out east (4,4)
17 Source of termites in garden consumed by anteater for one (8)
18 Dope — ecstasy, speed — available in form (8)
19 Rising Indian city getting in a state (7)

Across

1 Key worker's terribly tipsy after a short time (6)
4 Not on holiday, but having a ball (3,5)
10 Beach guard, perhaps, conveniently close to wave (9)
11 Maintain this will cover blemish (5)
12 Small sum of money received in post as payment (7)
13 Town working during, and after, strike (7)
14 Good-hearted libertine? (5)
15 Tree completely overshadows entrance to London thoroughfare (4,4)
18 Edible parts of fine leaves (5,3)
20 American soldier's sent back letter (5)
23 In title role, song performed when abandoned by love (7)
25 Manage stage production at empty theatre (7)
26 Sickly-looking party member (5)
27 Delay deciding no pay increase is to be given to 1 *ac* perhaps (9)
28 Players run between wickets (8)
29 Guy about to make a comeback, a tenacious sort (6)

Down

1 Succeeded in record attempt, hanging on the wall (8)
2 Placing ball on table, though not initially scoring at snooker? (7)
3 Plant in round container turning up in garden hut (9)
5 Whatever I do, it's onwards with energy with my autobiography (3,3,4,2,2)
6 Fail to declare weapon (5)
7 Official order's rewritten in dialect (7)
8 Man in charge keeps it flying (6)
9 Shifty soldier has hidden plan for screening apparatus (5,9)
16 Give false information about ship's prow sticking in mud near harbour (9)
17 Desert traveller has arrived, look (8)
19 Demanding fellow almost stuck in door (7)
21 Please start to give sanction (7)
22 Sever link, creating problem (4-2)
24 Taken to court, touring in Washington? (5)

64

65

Across

1 Crystal gazer originally taken in by 19 (3,5)
5 Boat's progress in tunnel (6)
8 Expose rook on its original square? Possibly (3)
9 Lectures go wrong in city (10)
10 Stir restlessly in wood that's ancient (8)
11 One spearing a kipper? (3-3)
12 Spiral galaxy, roselike at the edges (4)
14 He had a classical hangover cure (10)
17 In passing, frozen lump of earth seen across one curving path (10)
20 I can't return this river shrub (4)
23 Fetter great beast turned in by soldier (6)
24 Big animal circling round small cousin's place (8)
25 Love, according to a good girl, that seems to bring people closer (5-5)
26 Handbag regularly needed by woman (3)
27 Afterthought about one drink or another (6)
28 Leap made by old king around Fools' Day (8)

Down

1 Girl nearly gets to munch before toast can be made with it (9)
2 Asian native, one spotting sailor in sound (7)
3 Smoothly-flowing, on a Tornado's wings (6)
4 Rogue caught in stable with both hands grasping mare's tail (9)
5 Small light in fine harbour (7)
6 Hit upon brief answer where best friends might end up (9)
7 Through passport control, help European get round US tax authorities (7)
13 Get ahead of the rest on fast A1 (9)
15 Dodgy "hot" capital securing old banger (9)
16 Theft of pickled gherkins around area (9)
18 First lady's old man hiding options (7)
19 Mostly pretty girl that could hurt one, in the main (7)
21 City was elegant before now (7)
22 One making employer subservient to accountant (6)

65

66

Across

1 Stout Bobby takes you and me to court in the end (6)
4 Steal work of art (8)
10 Discover by chance what some answers do here (3,6)
11 Drink that's heated, say, when source of heat is round about (5)
12 New world dairymen get confused, in a parochial way (6-8)
14 Get to the point in the light (5)
16 Contrite competitor dropping out right after salesman (9)
18 Courteousness displayed by everyone in support (9)
20 French dish, or Indian? Just a penny in it (5)
21 Broadcast about state with climate control (3-11)
25 Supple man got down first (5)
26 King soon embraces intrepid maiden (9)
27 Gin — a type unusually from the Middle East (8)
28 Find a home in Yorkshire town (6)

Down

1 Champing at the bit, artillerymen circle republic (6,2,2)
2 Stupid mistake made by butcher? (5)
3 Help somebody gullible to be heard (7)
5 Bowled a wrong bowl (5)
6 Priest finally put away in Abbey (7)
7 Sponge given dough by financial backer (5-4)
8 Assignment — a skivvy set about it (4)
9 Bunk in which you'll see soldier go off (8)
13 After shuffling, ace, ten and, lastly, eight are all that turned up (10)
15 Friend vague about right way to see the future (9)
17 Caliph's worried about unknown examination (8)
19 Very old, about 101 being admitted (7)
20 Oriental crests seen all over the east (7)
22 Antelope in area between two American cities (5)
23 Books about a certain group of players (5)
24 Down in Oxford, it's dark (4)

66

67

Across

1 Composer's pen (4)
3 Stirring pap, Roman official holds a spatula (10)
9 Prime minister unlocked success? (7)
11 Surprise, no starter — time for little cake (7)
12 Shopping taking the blame in distortion of reality (6,7)
14 Sanctimonious joke about beautiful person (3-2)
15 Two short limbs to negotiate tempestuous conditions not on for swimmer (5,4)
17 Skinny? Back at university, gaining ounces getting drunk (9)
19 Unlimited jargon by thieves, primarily? (5)
21 Bird follows insect, one passing through bit of Leicester, perhaps? (8,5)
24 Insect eats fruit for food (7)
25 A card game requiring cut (7)
26 Early boat, among all the rest, that has many cheap offers (4,6)
27 Fine water source (4)

Down

1 Surge regularly times a hundred coming up to speed, reaching virtual capacity (10)
2 Setter's fiction elevated in spirit (7)
4 Etiquette unpolitic, strangely (9)
5 Machine no longer functioning round hospital (5)
6 Iran recruits a lunatic in preliminary to action (7-6)
7 Cry before killing friend in lap of god, almost (5-2)
8 Biblical character's gospel unopened (4)
10 Wallower almost breaking limb bearing into crowd? Beat it (8,5)
13 Parting note thanked originally at old-fashioned length (5,5)
16 Process food in wine cape in China (9)
18 Touchy and spiteful woman lifted hat (7)
20 Is it a hot surface? Good question! (7)
22 End of boot used in beating up teacher (5)
23 Is it above the ankle? A little lower (4)

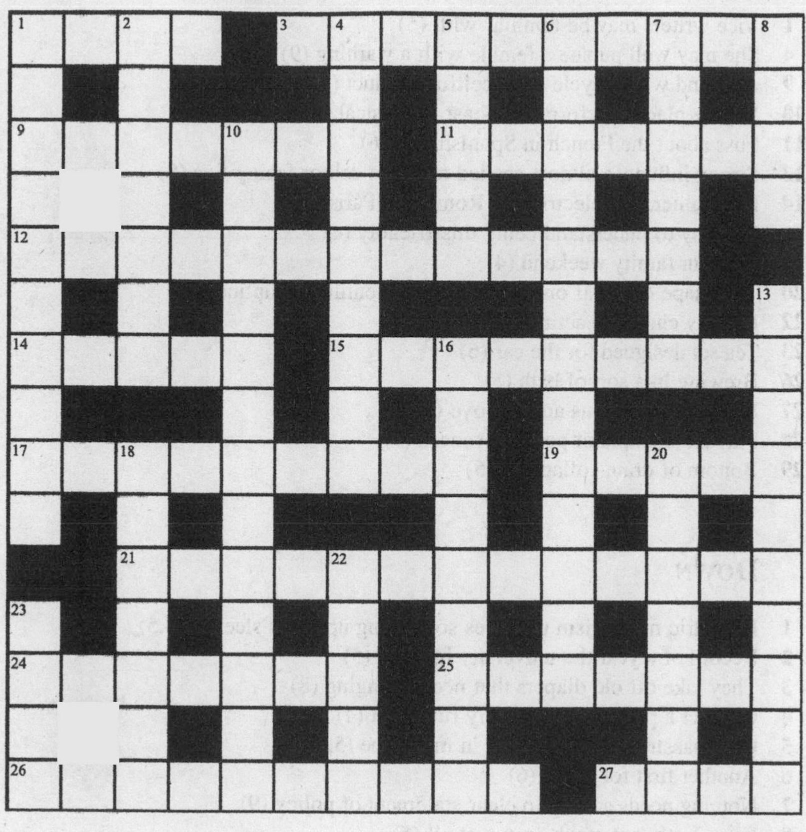

Across

1 Vice writers may be familiar with (5)
4 She may well pursue a female with a warning (9)
9 Abscond with a cycle — deceitful conduct (3-6)
10 Highly-placed performer's boast in musical piece (5)
11 Fuss about the French in Spanish city (6)
12 Strong influence almost needed to arrest debtor facing fine (8)
14 Entertainer, one electrifying Rome and Paris (10)
16 It's easy to understand being this friendly (4)
19 Hitch in family weekend (4)
20 The shape of a loaf once felt to be a scientific discipline (10)
22 Live by cheating, acting well (8)
23 Tea set designed for the car (6)
26 Brewing in a sort of bath (5)
27 Knock unconscious and remove (6,3)
28 Play an example of permissiveness (9)
29 Bottom of drain collapsing (5)

Down

1 Eccentric mannerism that uses something up one's sleeve? (4,5)
2 Record of a year the university ignored (5)
3 They take off old diapers that need changing (8)
4 Conflict a psychiatrist initially ruled out (4)
5 One maestro I'd write about in magazine (5,5)
6 Another first for Hugo (6)
7 Nothing needs adding to clear statement of policy (9)
8 From birth, not much grown at all (5)
13 Find and confirm (3,5,2)
15 Drinking vessel made from boxwood? (9)
17 Poor poet writing verse, or something worse (9)
18 Helpless because drink finally won (8)
21 Archetype of girl sailor (6)
22 Boot's paper tiger, for example (5)
24 Confessed, say, in a normal voice (5)
25 Not employed on the house (4)

68

69

ACROSS

1 Bony bits girl found in savoury dish (8)
5 Something on floor briefly left — bit of a flower (6)
10 Fish can be got with rod, that's very plain (9)
11 Nobody could have written this book (5)
12 Ruler, protected by escorts, arrived (4)
13 Mineral featured in note at back of old book (9)
15 Run — man is tired out (10)
17 Fatty stuff daughter's given to dog (4)
19 Transport system bringing objection around England's capital? (4)
20 Craftsman's pub somewhere off the M25 (7,3)
22 Kill time and order time to be killed (9)
24 Book succeeded after a little while (4)
26 Think of love, then waste away? (5)
27 European cleaner I'd sacked (9)
28 It could be a donkey that's put to work (6)
29 Character being grabbed and cuddled (8)

DOWN

1 The old man needs two assistants in the office (4)
2 Suffer in the ring but be favourite to win? (4,4,7)
3 Twine hanging under middle of blue cloth (8)
4 Panic in the manner of the Marines (5)
6 Make off with sailor going over the Channel (6)
7 Man spellbound — it could be the light-haired girls (8,7)
8 What editors write about Scottish town's preachers (3,7)
9 Make an initial modification to flush sewage (8)
14 Struggle — not a love game (10)
16 Irregular disco rap for dancing (8)
18 Animal needing to have a meal, tucking into fruit (5,3)
21 Given a very long time, unlikely flier can become one (6)
23 Spanish girl: feel her near yet oddly missing! (5)
25 Brought up short in race, doing somersault (4)

69

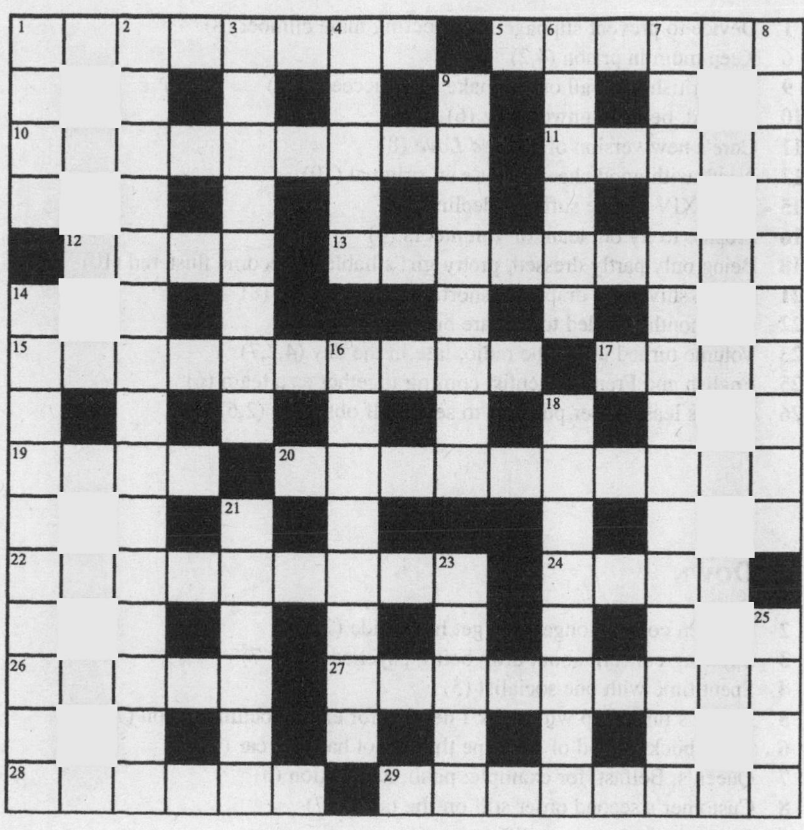

Across

1 Device to prevent slippage is protecting male climber (8)
6 Keep mum in prison (4,2)
9 The British went all out to make this succeed (7,6)
10 Banned, because unwelcome (6)
11 Cure's new version of *Tainted Love* (8)
13 A pub with good cheer? That's an original (10)
15 Louis XIV maybe suffered decline (4)
16 Prepare to fly out team of volunteers (4)
18 Being only partly dressed, pretty girl's liable to become flustered (10)
21 I go in shivering, draped in short, jaunty bathrobe (8)
22 Two months needed to prepare herbal extract (6)
23 Volume turned up on the radio, late in the day (4,2,7)
25 English and French scientist coming together as a team (6)
26 Stamps leaseholder put next to seal, half obscured (2,6)

Down

2 Chicken cooked longer will get hot inside (7)
3 Showing concern about drug being psychedelic (4-7)
4 Spent time with one socialist (5)
5 Friend's turned up with book I needed for exam documentation (7)
6 Marshbuck: a kind of antelope that is not hard to rear (9)
7 Queen's, Belfast, for example: posh, its location (3)
8 Customer's second order still on the table? (7)
12 Grade that can make a difference (11)
14 Inspection copy includes old record (9)
17 Still a tortuous climb to get round peak of Everest (7)
19 Flashy lights in street South York rigged up (7)
20 Hannibal takes new stand (7)
22 Group of splendid stars performed in days of yore (5)
24 Blade slicing head off pig (3)

71

Across

1 Mixed drink peeler's garnished with a slice (7)
5 Soda dispenser making short drink sweet (6)
8 Distance covered by boy with appeal over maximum duration (4,5)
9 Laceration occurs here? (5)
11 Hysterical clamour about extreme letters (5)
12 Being gutted, one loses it during English exam? (9)
13 Work old jock gets in any other business — lucky break? (1,4,3)
15 Maiden I love on rotating platform (3,3)
17 Last month, extremely unhelpful skip overturned (6)
19 Person with tobacco addiction may get it to stop abruptly (5,3)
22 Pester garage, wrongly concealing tax (9)
23 Primarily guttural and coarse-sounding? (5)
24 Shows impatience with one Rwandan group (5)
25 David's wife cleanses Hebrew area (9)
26 Knock back pounds in ice cream, all together (2,4)
27 Felt deeply about son getting settled (7)

Down

1 Article shifting a clot in a gut? (13)
2 Harpsichord or cello with central part changed, to a commercial degree (7)
3 Prosperous friend of the speaker (5)
4 Once leading Frenchman and Englishman abroad, irrational and almost sullen (8)
5 Outsiders from Shiloh take tribal Jewish settlement (6)
6 Quietly, Dickens arranged to lose daughter, following his hypocritical character (9)
7 Working to rule, I will be an isolated person (7)
10 Destined for Forth's borders, conveyed by sled with dexterity (7-2-4)
14 Half-heartedly condemn man thrown to lions in terrible den (9)
16 Needle trendy judge with 'arangue? (8)
18 Tense time before note's twice removed from this number (7)
20 Wealthy circle whine over conservationists (7)
21 One chapter underpinning the core of William Blake's verse (6)
23 Theatre cat curtailed male one's ardour (5)

71

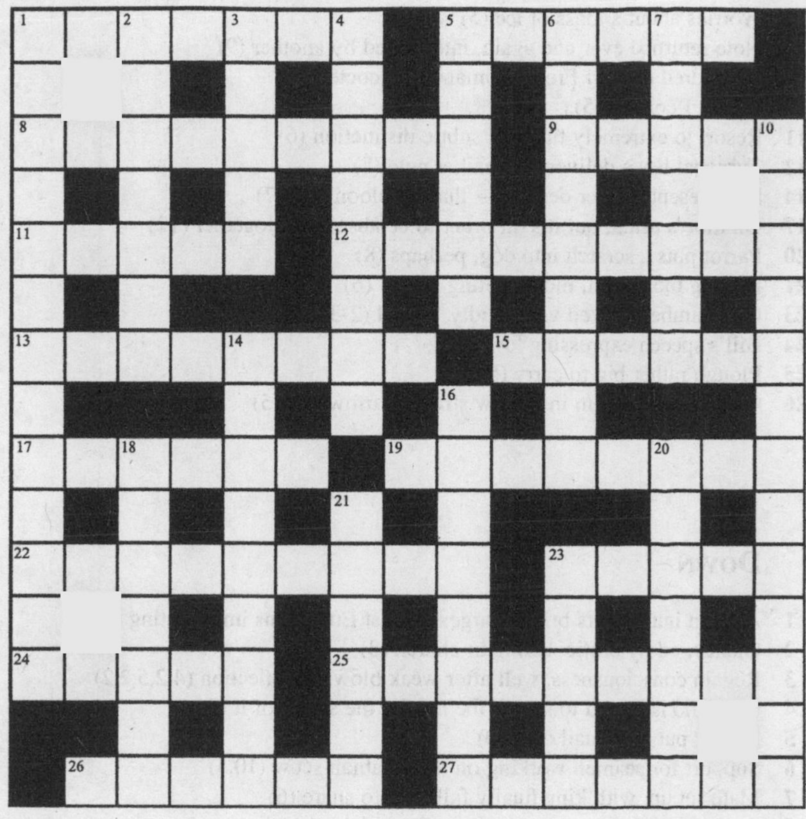

72

Across

1 Worries about a mass of ice (5)
4 Note returned ever and again, interrupted by another (9)
9 A hundred tales of French romance concocted (9)
10 In part I convey (5)
11 Resort to extremely tiny and subtle distinction (6)
12 Turncoat has a delivery of mail, a note (8)
14 Bill presented after dessert — that's a bloomer (5,7)
17 On time's usual, but it's different to be absolutely together (12)
20 Parrot puts a scratch into dog, perhaps (8)
21 Feeling more pain, hide greeting in tree (6)
23 One number, played very loudly, is foul (2-3)
24 Bill's speech expressing love (9)
25 Plough rather big to carry (5,4)
26 Colour starts to run in narrow streaks earthwards (5)

Down

1 African inhabitants bring charges against Europeans immigrating (8)
2 Glance, oddly erotic, taking in church (8)
3 Regain consciousness, well after weak blow, on reflection (4,2,5,2,2)
4 One who is bound to search the net, by the sound of it (4)
5 Doctor put an e-mail out (10)
6 Support for seamen working on this Arabian scow (10,5)
7 Mate set up, with king finally falling into snare (6)
8 Avoid points to argue over (6)
13 Strive to capture ram — scold and curse (10)
15 South American girl finds accommodation in the outskirts of Brighton (8)
16 Alienate English oddball (8)
18 Start season well (6)
19 Group of trees in a ring? On the contrary, a rut (6)
22 Noise from farmyard goes over river to upland (4)

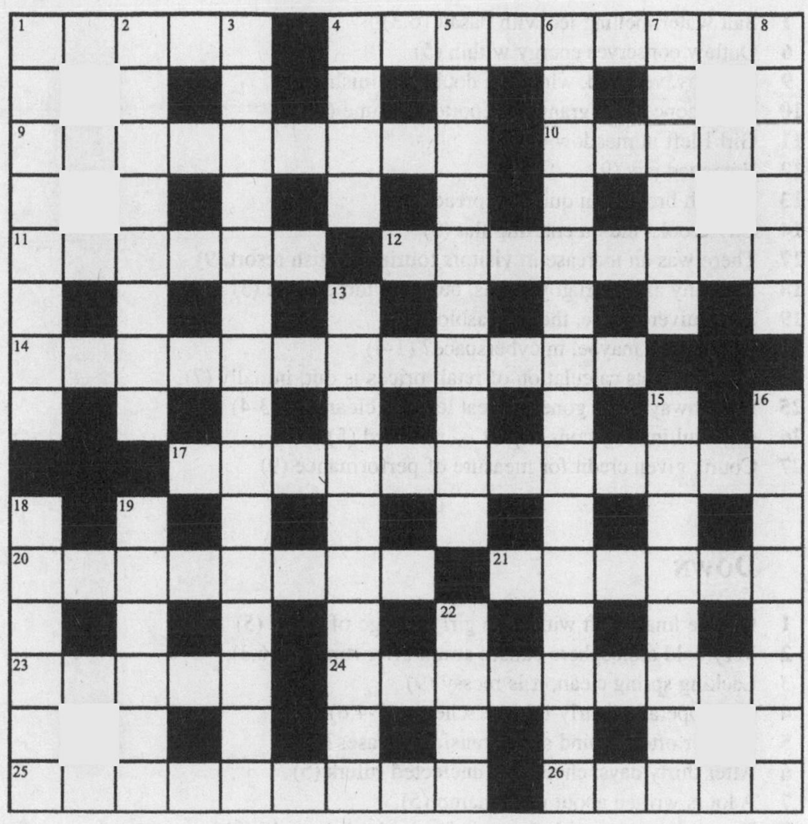

73

Across

1 Salt water melting ice with basalt (6,3)
6 Outlaw conserves energy within (5)
9 Cowboy, very old, without a doubt diminished (7)
10 Wrap concealing granny's reported fortune (7)
11 Girl I left in meadow (5)
12 Extended jaw (9)
13 English brought in quickly spread (5)
14 City crooks hid on entering flat (9)
17 There was an increase in visitors touring British resort (9)
18 As many as four rugby teams, but only three score (5)
19 An anniversary tie, then in fashion (9)
22 Kidnapped, maybe, in cyberspace? (1-4)
24 Sign suggests calculation of retail prices is out, initially (7)
25 Icy runway one's gone to great lengths clearing? (3-4)
26 Cheat ultimately cons expert – not hard (5)
27 Courts given credit for measure of performance (9)

Down

1 George finally left with little girl on edge of wood (5)
2 Very cold atmosphere causes strain after running (6,3)
3 Lacking spring clean, it is messy (9)
4 Cells operating early-release scheme (5-4,6)
5 Lawyer often behind swift transfer of cases? (9-6)
6 After thirty days, cheers for unelected rulers (5)
7 A lot is written about new champ (5)
8 For men meeting girl, flowers simply don't have it! (9)
13 Advocates of her achievements are repulsed by short skirts (9)
15 Sounded disapproving while talking over boy's great glee (9)
16 It's common thinking in Latin (3,6)
20 Girl's taken up complaint with one (5)
21 Stone used to bump off a leader in Zanzibar (5)
23 Kiosk regularly lined with a soft stuffing (5)

73

74

ACROSS

1 Risk in shooting costly film involving small company (8)
6 Chesterton's wicked tradesman's more bloated, it's said (6)
9 Quick course securing professional employment (6)
10 Stupid, unlike this puzzle! (8)
11 Tooth remnant surgeon finally removed from animal painter (4)
12 Report of horse-breaker's delaying tactics (10)
14 Creature a king's given by everyone in court (8)
16 Individual welcomed in Madeira (4)
18 Ten characters in a Berkshire town (4)
19 One-time players making extortionate demands? (8)
21 A foreign maiden Lucifer destroyed, being so cruel (10)
22 Agreeable place near Cagnes-sur-Mer (4)
24 Cooking-pot foretopman Budd is able to assemble (8)
26 Do porridge, and make light of it (6)
27 Marketplace in Athens stocking new cloth (6)
28 Giving up and giving way (8)

DOWN

2 Not proper, to doze during Computer Studies (5)
3 Come a cropper with regard to surplus (11)
4 Cow given drug appropriate to animals kept outside (8)
5 Delight king found in free elections? Goodness! (6,4,5)
6 Dirty American town's exalted past (6)
7 Bird of prey's cry spoken of by 8 (3)
8 Citizen English regarded as sympathetic (4,5)
13 Defective vehicle overturned on model (11)
15 Order that should be paid to a teller's account? (9)
17 Like schooners fit for marketing, say (8)
20 Shrub about to be planted in a cold American state (6)
23 Exult over new sovereign or old coin (5)
25 See visited by ecclesiastical leader, one of several popes (3)

74

75

ACROSS

1 Temporary worker's sloppy habits? (7)
5 Limb: part of the body to be kept in shape (7)
9 No answer in the city (3)
10 Made to stop, having got weary and depressed (7,4)
11 The whole thing's about to fall in (8)
12 Guy holding good — last bit of rope in camp (6)
15 Goodbye, girl (4)
16 Revive old railways in an area of land (5,5)
18 We're told to take gift: cheap trinket, perhaps (5-5)
19 A doctor's book (4)
22 Insects husband squashed in journey south (6)
23 Clear soup's OK to keep in summer month (8)
25 I take hash with GI, stupidly getting stoned (4,2,1,4)
27 Strong feeling regularly displayed in Israel (3)
28 Leader's giving power to soldiers during fighting (7)
29 Slope leading from top of foreign trail (7)

DOWN

1 Change name in secret (7)
2 Breed of dog still seen in most of Scottish islands, right? (4,7)
3 A business in sound state (6)
4 Count comes in to celebrate a bit (10)
5 Confused heartless old buffer (4)
6 Bird signalling rain later in the day? (8)
7 A teacher's endless trouble (3)
8 Brother briefly observed rolling in it (7)
13 Attack nursemaid developing love of good food (11)
14 Empty container — coin due to be chucked across it (10)
17 Aircraft captain is on course (3-5)
18 The deck's not initially refurbished in these sailing vessels (7)
20 Swedish physicist always to be found in location out east (7)
21 Not all different birds (6)
24 Love to whip up a quick dish (4)
26 It's torture for college employee (3)

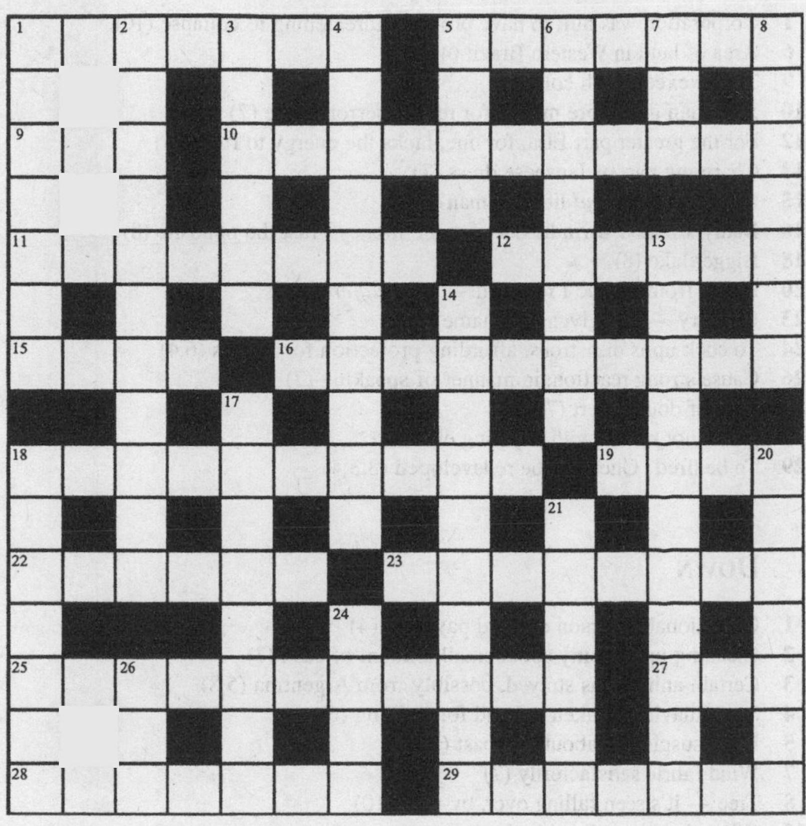

76

ACROSS

1 Corporation was hurt to have property threatening to collapse (10)
6 Area of land in Western Brazil (4)
9 Turner excels with colour (7)
10 Salesman gets more money for repeat performance (7)
12 For the greater part Elsa, for one, lacks the energy to run (5,5)
13 Charming part of Japanese dress (3)
15 Marrying large *and* little woman? (6)
16 Badly affected, swindle out of some money? Just the opposite (8)
18 Bigger lake (8)
20 Exiles from France I returned — *c'est différent* (6)
23 Hostelry — one given new name (3)
24 To cock up is disastrous, affording protection for insects (6,4)
26 Cause strong reaction, in manner of speaking (7)
27 Face of dour expert (7)
28 Cattle not mixed with anything else (4)
29 To be fired? Ghetto to be redeveloped (3,3,4)

DOWN

1 Objectionable person delayed payment (4)
2 Cleaning up, gloomy about small amount of cash (7)
3 Certain animal has strayed, possibly from Argentina (5,8)
4 Scandinavian is taken in hand for training (6)
5 Bond suspicious about bombast (8)
7 Wind fabric satisfactorily (7)
8 Tree — it's seen falling over, by me? (10)
11 Gilbert's object all sublime
Put in rime and only rime (6,7)
14 Gorge with popular chap from East Africa (10)
17 Specific cape on island (8)
19 Father holds one of the highest cards — that'll solve everything (7)
21 Port with good taste in the past (7)
22 Purchased some wood, a ton (6)
25 Suitable assembly (4)

76

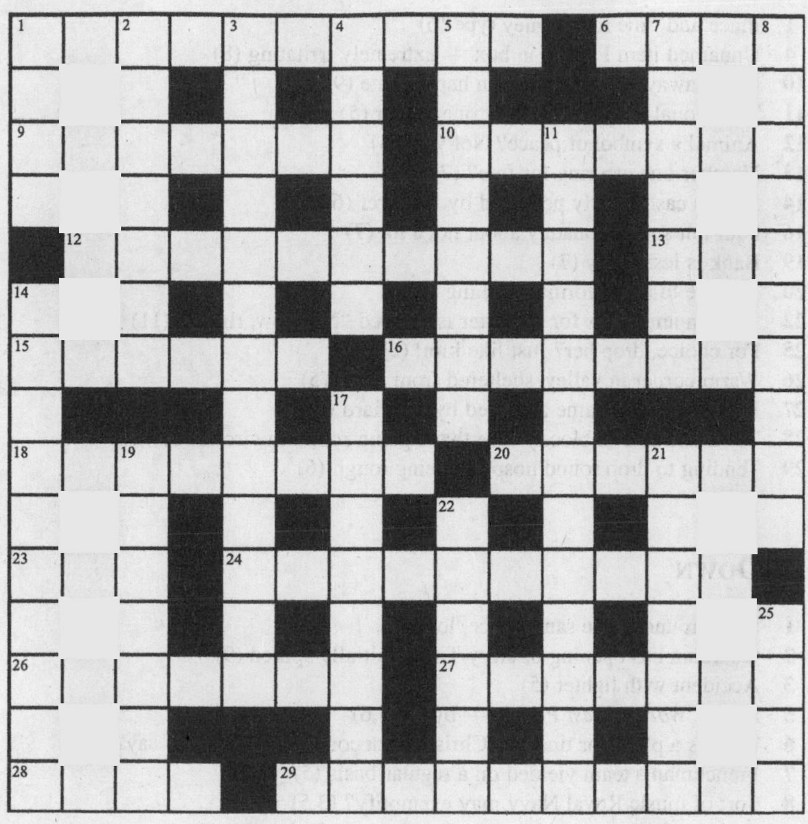

77

Across

1 Place and time for plaguey type (6)
4 Unnamed item I found in box — extremely irritating (8)
10 Hurry away, due to change in happy state (9)
11 Traditional knowledge about one flower (5)
12 Animal a symbol of peace? Not very (3)
13 Number one can ring for free? (7,4)
14 A little cash quietly pocketed by manager (6)
16 Paul felt so passionately about not a lot (7)
19 Bank is less misty (7)
20 Decline to act in formal clothing (2,4)
22 Woman ambitious for daughter is ordered "Not now, right?" (11)
25 For choice, drop her? Just like him! (3)
26 Warm corner in valley, sheltered from north (5)
27 Irish husband's name accepted by Spaniard (9)
28 Try a criminal in bloody case that's going round in circles (8)
29 Tending to drop round hospital, being rough (6)

Down

1 Drive round in the same place, look (6)
2 Guardian has opening of story Times initially spiked (9)
3 Accident with lighter (5)
5 Novel: *World's Best Pistols* — by? (3,5,6)
6 There's a place for tinsel as Christmas decoration, would you say? (9)
7 Frenchman's team yielded on a regular basis (5)
8 Sort of music Royal Navy may exemplify? (3,5)
9 Less sensitive person doing the grind, one calculates (6,8)
15 Bow to people covered in part in sheepskin (9)
17 Object about obnoxious type getting in the way (9)
18 Farm animal given first aid looking unhappy? (8)
21 Humming, as way's dark (6)
23 Chap as oarsman losing weight — good to replace it (5)
24 He let America go in one direction (5)

77

78

Across

1 He may be looked down on by the girl he woos (9)
6 A prison officer's salary (5)
9 Country house appears posh, after a couple of drinks (7)
10 Feature encompassing rejection of unsatisfactory dress material (7)
11 It's not effervescing, for all that (5)
12 Outward features unknown in immortal saint (9)
14 Ben's neighbour? On the contrary (3)
15 Blooming radiation emission not involving uranium! (11)
17 Ounce of powder a son spilt crossing lake (4,7)
19 Part of theatre that's mine (3)
20 Solicit loan from person getting work in pool? (5-4)
22 Foreign character in British reserve force (5)
24 List is examined at first by cohabiting couple (7)
26 Hard to get into revised score incorporating unknown movement (7)
27 Bold fellow digesting Eliot, initially (5)
28 Disliked having a French university in a London district (9)

Down

1 Plunders wines (5)
2 Writer like Balzac, perceptibly on a roll (7)
3 Quantity of thread required to secure borders of lace (9)
4 Book identifying old cat? (11)
5 Legendary flyer takes up French horn (3)
6 Starting at nursery level, one goes downhill fast (5)
7 Burden, having brother in control (7)
8 Dishevelled players showed distress (9)
13 Springer's been active? Then all is revealed (3,4,2,2)
14 Surrounding creature that's protecting burrow (9)
16 Position of trainee serving time in rebel leader's joint (9)
18 Mostly work fast to get rich (7)
19 Like serous membrane? More than one, it's said (7)
21 Dangerous description of Jacob's brother Esau (5)
23 Disheartened general leaves beauty for illicit liaison (5)
25 Native Australian depositing millions in some Western states (3)

79

Across

1 Sort of spin from Labour leader, say, before recess (3,5)
5 Rest needed by chap going to pot (6)
9 He did paintings found in bottom drawer after short time (8)
10 Cut short with revolutionary saw, perhaps, used too much (6)
12 Paper I deplore condemned perfect arrangement (5-3,5)
15 Lines of verse delivered by Philoctetes (5)
16 Be extremely excited by game with hands (2,7)
17 Quisling was a real pest (6,3)
19 China from Valencia? (5)
20 In psychological assessment chart, the cross is misplaced (9,4)
22 Man due to crash in far from tidy bed like this (6)
23 *The Sting* is both a film and play (4,4)
25 Sycophants longing to get out of southern state (3-3)
26 Urgent way travel book's been written (8)

Down

1 Title of a person at end of Orlando novel (5,5)
2 Not even genius could produce such animals (3)
3 Admiration for relation (7)
4 Hymn bringing to a mass a vitality and beauty (7,5)
6 Heroic knight or distressed young man? (7)
7 Record it in a new version? That's not practical (11)
8 Stagger out of some film (4)
11 Flyer, one offering lift, interrupts one in orbit — an offensive habit (6,6)
13 Opera house built up following after favourite monarchs came together (5,6)
14 When kind people start off thoughtful collection (10)
18 A new partner receiving good cut (7)
19 Ghanaians emphatically refuse to go into capital (7)
21 Landing made by legend on the radio (4)
24 Not fine to confront champion (3)

79

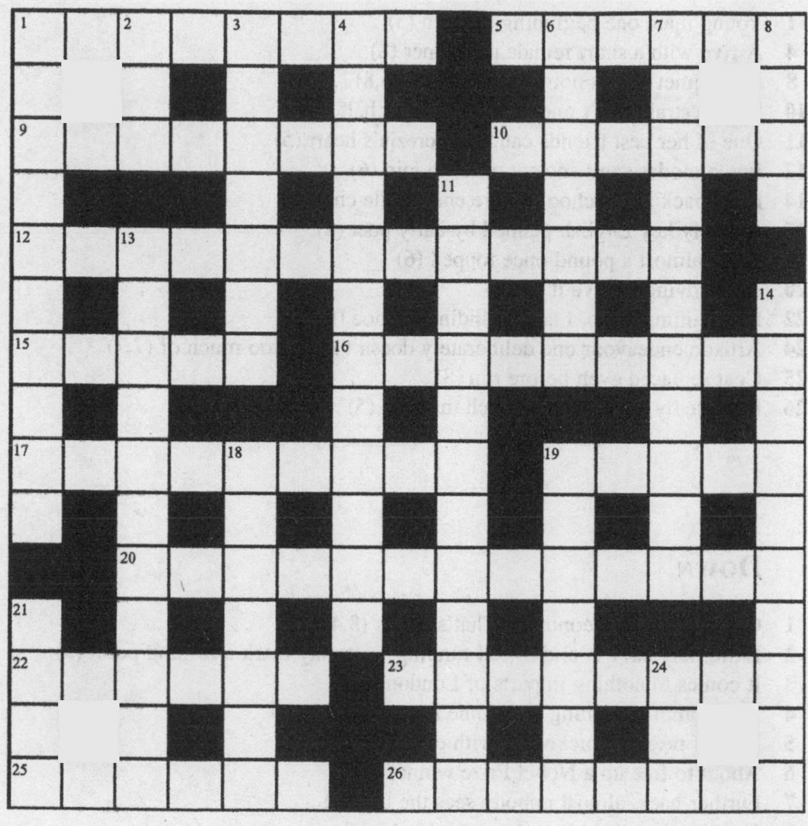

Across

1 Young man, one beginning to learn (5)
4 Arrive with a short female performer (8)
8 Dead quiet? Most not wanting a fuss (6,8)
10 Due to strain, isn't one jumpy too? Not half (9)
11 One of her best friends caught Lucrezia's heart (5)
12 Force modern and ancient cities to join (6)
14 Rush back after school before end of idle chat (8)
17 Possibly key English penned by early poet (8)
18 Fruit, almost a pound once topped (6)
20 One's dying to give it up (5)
22 Little animal, also, I caught hiding in shoe (9)
24 Artistic endeavour one deliberately doesn't make too much of (7,7)
25 Coat replaced even before run (8)
26 Reportedly influenced by such material (5)

Down

1 Certain to clinch contract? That's a plus (8,4)
2 Extremely pacy at end of fell running — it may mark a turning point (5)
3 It comes to nothing in parts of London (9)
4 Join compiler coming into some money (6)
5 Pots of preserve picked up with cola I ordered (8)
6 About to free up a Nobel Prize winner (5)
7 Further back, almost nobody sees the PM (9)
9 Very large bird in short extract from American magazine (4,8)
13 Men like monk's singing (9)
15 Island where one missing flight disappeared (9)
16 Covering up, doctor to probe injury — prod heartily! (8)
19 Girl getting cross over referee on the right wing (6)
21 Initially tussled with italicised clues encountered in *Times Two*? (5)
23 Poem about soccer team in compound (5)

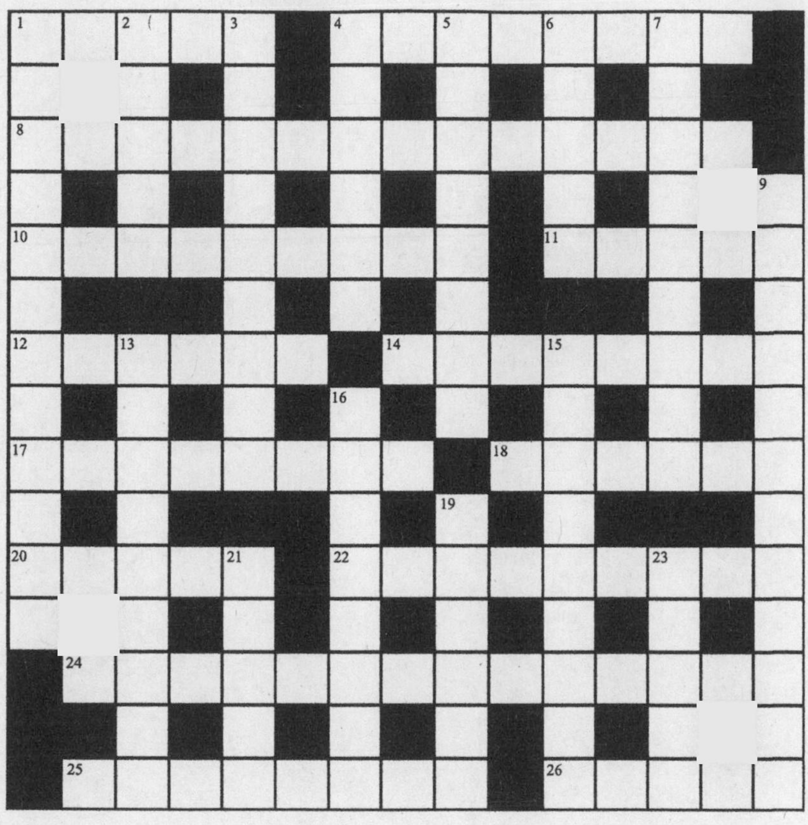

THE SOLUTIONS

1

C	A	R	O	L		I	M	P	R	E	C	I	S	E
O		H		I		B		R		P		N		V
M	O	O	N	S	H	I	N	E		I	N	A	N	E
E		N		S		S		V		T		M		R
I	C	E	B	O	X		T	A	X	O	N	O	M	Y
N				M		P		L		M		R		N
T	I	T	L	E	P	A	G	E		E	R	A	T	O
O		E				W		N				T		W
O	S	S	I	A		N	I	C	O	T	I	A	N	A
N		T		B		B		E		R				N
E	M	I	S	S	A	R	Y		F	I	T	T	E	D
S		F		I		O		S		R		O		T
O	R	I	O	N		K	I	T	T	E	N	I	S	H
W		E		T		E		A		M		L		E
N	O	R	T	H	E	R	L	Y		E	S	S	E	N

2

T	W	A	D	D	L	E		C	A	S	S	A	V	A
O		C		R		V		I		H		S		R
R	A	C	K	E	T	E	E	R		O	P	I	U	M
T		U		A		N		C		R		N		A
		S	I	D	E	S	P	L	I	T	T	I	N	G
C		A				O		E		L		N		E
H	E	L	L	B	E	N	T		B	I	G	E	N	D
A				A		G		L		S				D
R	E	P	O	R	T		S	A	N	T	I	A	G	O
I		A		C		P		V				U		N
T	H	R	E	E	D	A	Y	E	V	E	N	T		
A		S		L		R		N		M		O		G
B	I	N	G	O		K	I	D	N	A	P	P	E	R
L		I		N		I		E		I		S		I
E	X	P	L	A	I	N		R	E	L	A	Y	E	D

3

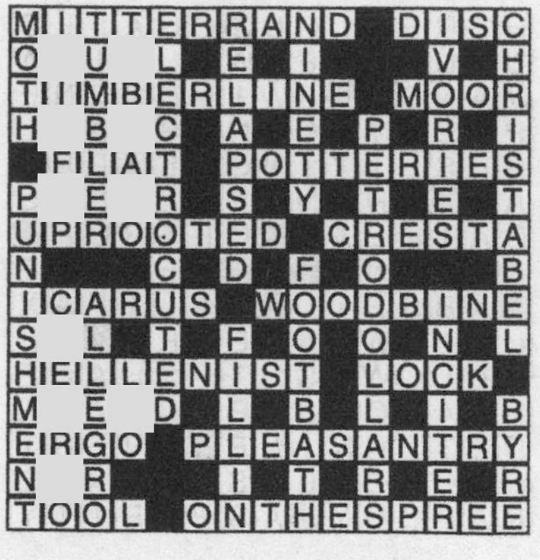

M	I	T	T	E	R	R	A	N	D		D	I	S	C
O		U		L		E		I				V		H
T	I	M	B	E	R	L	I	N	E		M	O	O	R
H		B		C		A		E		P		R		I
	F	L	A	T		P	O	T	T	E	R	I	E	S
P		E		R		S		Y		T		E		T
U	P	R	O	O	T	E	D		C	R	E	S	T	A
N				C		D		F		O				B
I	C	A	R	U	S		W	O	O	D	B	I	N	E
S		L		T		F		O		O		N		L
H	E	L	L	E	N	I	S	T		L	O	C	K	
M		E		D		L		B		L		I		B
E	R	G	O		P	L	E	A	S	A	N	T	R	Y
N		R				I		T		R		E		R
T	O	O	L		O	N	T	H	E	S	P	R	E	E

4

W	I	N	D		B	O	O	K	M	A	R	K	E	D
	G		E		I		N		I			Y		E
K	N	O	C	K	K	N	E	E	D		F	L	A	K
	O		E		I		P		T			I		K
T	R	U	M	A	N		A	B	E	D	N	E	G	O
	A		V		I		R		R				U	
I	M	P	I		S	P	E	R	M	A	C	E	T	I
	U		R				N				O		T	
A	S	P	I	D	I	S	T	R	A		N	E	E	P
	E				C		F		N		F		R	
A	S	Y	N	D	E	T	A		O	B	L	A	S	T
G		A			F		M		T		U		N	
O	C	H	E		A	R	I	T	H	M	E	T	I	C
G		O			L		L		E		N		P	
O	Z	O	N	E	L	A	Y	E	R		T	R	E	K

SOLUTIONS

5

E	X	C	E	P	T		B	R	A	S	S	H	A	T
V		A		E		B		I		C		E		O
I	N	T	E	R	B	R	E	D		R	E	A	D	E
D		W		C		I		I		U		R		C
E	R	A	S	U	R	E		N	E	M	E	S	I	A
N		L		S		F		G				E		P
C	A	K	E	S		E	M	B	O	L	I	S	M	
E				E		N		R		O				I
	M	A	S	S	A	C	R	E		V	A	L	I	D
E		T				O		E		I		E		O
M	A	R	A	B	O	U		C	O	N	G	E	A	L
B		O		E		N		H		G		W		A
A	L	P	H	A		T	R	E	N	C	H	A	N	T
R		H		N		E		S		U		R		E
K	E	Y	B	O	A	R	D		S	P	I	D	E	R

6

A	G	O	U	T	I		B	U	T	T	O	N	U	P
	R		N		N		U		O		B		R	
M	A	N	D	A	T	O	R		W	H	I	T	B	Y
	V		E		H		N		E				A	
D	E	G	R	E	E		O	I	L	S	T	O	N	E
	S		S		K		U				H		I	
			C	O	N	S	T	I	T	U	E	N	T	S
	W		O		O				A		O		E	
S	H	I	R	T	W	A	I	S	T	E	R			
	E		E				L		T		E		U	
H	A	N	D	B	I	L	L		O	U	T	I	N	G
	T				N		I		O		I		S	
S	E	N	L	A	C		C	L	I	N	C	H	E	S
	A		E		A		I		S		A		E	
D	R	A	G	O	N	E	T		T	A	L	E	N	T

7

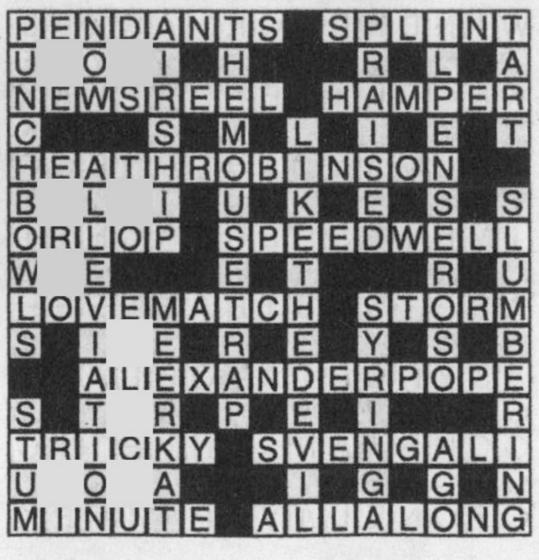

P	E	N	D	A	N	T	S		S	P	L	I	N	T
U		O		I		H				R		L		A
N	E	W	S	R	E	E	L		H	A	M	P	E	R
C				S		M		L		I		E		T
H	E	A	T	H	R	O	B	I	N	S	O	N		
B		L		I		U		K		E		S		S
O	R	L	O	P		S	P	E	E	D	W	E	L	L
W		E				E		T				R		U
L	O	V	E	M	A	T	C	H		S	T	O	R	M
S		I		E		R		E		Y		S		B
		A	L	E	X	A	N	D	E	R	P	O	P	E
S		T		R		P		E		I				R
T	R	I	C	K	Y		S	V	E	N	G	A	L	I
U		O		A				I		G		G		N
M	I	N	U	T	E		A	L	L	A	L	O	N	G

8

H	E	R	E	F	O	R	D		C	H	A	F	E	D
U		U		R		E		W		U		I		W
M	A	N	G	O		S	W	I	N	B	U	R	N	E
P		E		M		C		N		B		E		L
B	E	S	E	T	T	I	N	G		L	A	P	E	L
A				H		N		E		E		R		S
C	A	T	H	E	A	D		D	E	T	R	O	P	
K		A		N						E		O		E
	A	B	S	E	I	L		H	A	L	I	F	A	X
H		U		W		A		E		E				P
A	L	L	O	W		S	H	I	P	S	H	A	P	E
R		A		O		S		R		C		Z		R
D	E	T	E	R	M	I	N	E		O	V	E	R	T
U		O		L		E		S		P		R		L
P	A	R	O	D	Y		A	S	P	E	R	I	T	Y

9

F	R	I	C	T	I	O	N		C	O	B	W	E	B
A		N		A		L				N		H		E
B	I	S		L	A	D	Y	M	A	C	B	E	T	H
R		I		I		W		A		E		E		E
I	N	D	I	S	T	I	N	C	T		A	D	A	M
C		E		M		V		K		C		L		O
	C	R	E	A	S	E		E	A	R	N	E	S	T
H				N		S		R		U				H
O	C	U	L	I	S	T		E	N	S	I	G	N	
M		N		C		A		L		T		A		A
E	M	I	R		I	L	L	S	T	A	R	R	E	D
S		F		O		E		H		C		B		D
P	L	O	U	G	H	S	H	A	R	E		A	I	L
U		R		R				R		A		G		E
N	U	M	B	E	R		S	K	I	N	H	E	A	D

10

H	A	L	L	M	A	R	K	S		S	O	W	E	R
A		I		I		I		A		W		H		A
R	E	V	E	L		M	A	M	M	A	L	I	A	N
D		E		L		B		A		R		P		K
C	A	N	A	S	T	A		R	E	M	O	R	S	E
O				T		U		I				O		R
P	R	I	M	O	R	D	I	A	L	S	O	U	P	
Y		N		N						H		N		P
	A	D	V	E	N	T	C	A	L	E	N	D	A	R
W		E				I		R		L				U
H	E	X	A	G	O	N		B	E	L	A	T	E	D
I		L		O		T		I		F		O		E
S	T	I	N	G	I	E	S	T		I	N	K	I	N
K		N		O		R		E		S		Y		C
Y	O	K	E	L		N	O	R	T	H	P	O	L	E

11

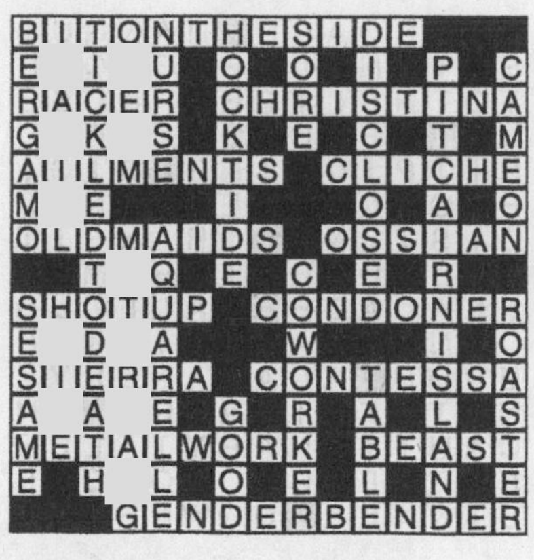

B	I	T	O	N	T	H	E	S	I	D	E			
E		I		U		O		O		I		P		C
R	A	C	E	R		C	H	R	I	S	T	I	N	A
G		K		S		K		E		C		T		M
A	I	L	M	E	N	T	S		C	L	I	C	H	E
M		E				I				O		A		O
O	L	D	M	A	I	D	S		O	S	S	I	A	N
		T		Q		E		C		E		R		
S	H	O	T	U	P		C	O	N	D	O	N	E	R
E		D		A				W				I		O
S	I	E	R	R	A		C	O	N	T	E	S	S	A
A		A		E		G		R		A		L		S
M	E	T	A	L	W	O	R	K		B	E	A	S	T
E		H		L		O		E		L		N		E
			G	E	N	D	E	R	B	E	N	D	E	R

12

E	X	P	O	S	U	R	E		H	E	R	M	E	S
M		A		O		O				F		E		O
P	O	R	T	F	O	L	I	O		F	I	N	A	L
I		I		T		L		N		A		D		S
R	O	S	E	T	T	E		E	N	C	H	A	N	T
E				O		R		F		E		C		I
		H	Y	P	O	C	H	O	N	D	R	I	A	C
C		O				O		R				T		E
H	A	L	F	H	E	A	R	T	E	D	L	Y		
A		D		O		S		H		I				C
T	O	W	B	O	A	T		E	R	O	T	I	C	A
T		A		D		E		R		C		D		N
E	X	T	O	L		R	H	O	D	E	S	I	A	N
L		E		U				A		S		O		E
S	C	R	U	M	P		A	D	V	E	R	T	E	D

SOLUTIONS

13

```
GREASE#SUPERMAN
O#V#U#C#P#A#I#E
GLASSBLOWER#SIX
E###A#A#A#P#C#T
TETANUS#ROLLOUT
T#A#N#S#D#U#N#O
EPIDEMIOLOGIST#
R#C###C#Y###T#R
#PHANTASMAGORIA
C#I#I#L#O#R#U#S
RECITAL#BLAZEUP
U#H#P#A#I#N###U
TAU#INTELLIGENT
C#A#C#I#E#T#N#I
HONGKONG#GARDEN
```

14

```
##CAPITALGAINS#
F#A#O#R#O#I#A##
REMORSE#GIRASOL
E#B#L#A#I#C#T#E
NERVOUS#CARDIFF
C#I#C#U###E#E#T
HACEK#ROADWORKS
P#####E#L#####T
OUTSOURCE#ALLOA
L#O#R###X#R#A#N
IMPUGNS#ATTIRED
S#I#A#T#N#I#G#I
HACKNEY#DISCERN
##A#I#E#E#A#S#G
#ALICESPRINGS##
```

15

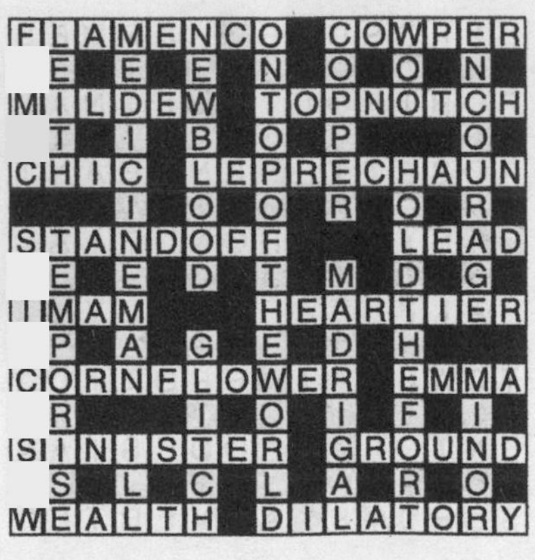

16

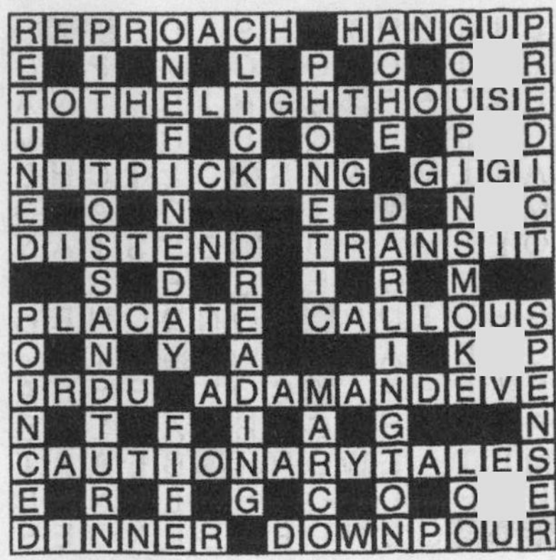

17

M	I	G	H	T		P	A	S	S	E	R	S	B	Y
U		O		E		S		E		Y		T		E
S	H	A	M	A	T	E	U	R		E	M	A	I	L
I				U		U		G		S		G		P
C	A	M	B	R	I	D	G	E	B	L	U	E		
O		A		N		O				E		F		C
L	A	S	T		A	N	D	S	O	F	O	R	T	H
O		T		S		Y		A		T		I		I
G	U	E	S	T	I	M	A	T	E		O	G	R	E
Y		R		E				E		B		H		F
		C	A	P	I	T	A	L	L	E	T	T	E	R
A		L		D		E		L		H				A
P	R	A	D	O		S	P	I	D	E	R	W	E	B
E		S		W		T		T		L		O		B
D	I	S	E	N	G	A	G	E		D	H	O	T	I

18

O	R	C	H	A	R	D		S	P	A	R	R	O	W
N		H		P		O		P		P		E		A
S	T	A	M	P	D	U	T	Y		P	A	P	E	R
I		L		R		B				A		L		R
D	E	L	I	A		L	I	T	T	L	E	E	V	A
E		A		I		E		R				T		N
	T	H	E	S	E	C	R	E	T	A	G	E	N	T
S				E		R		E		G				Y
T	R	A	P	D	O	O	R	S	P	I	D	E	R	
O		L				S		U		T		N		S
W	O	M	A	N	I	S	E	R		A	I	T	C	H
A		O		A				G		T		R		U
W	A	N	E	D		P	R	E	S	I	D	E	N	T
A		E		I		O		O		O		A		U
Y	A	R	D	A	G	E		N	O	N	S	T	O	P

19

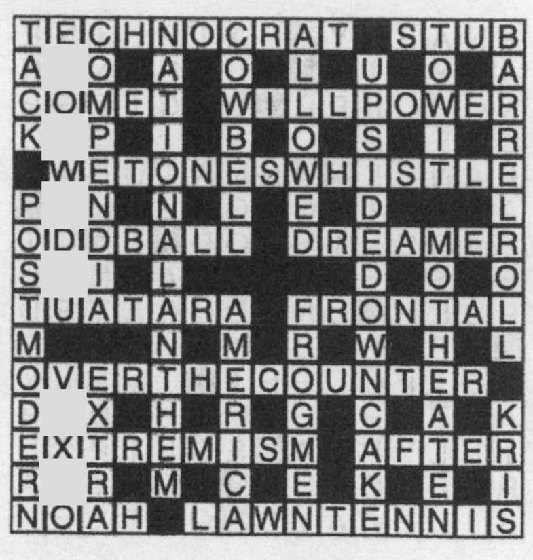

20

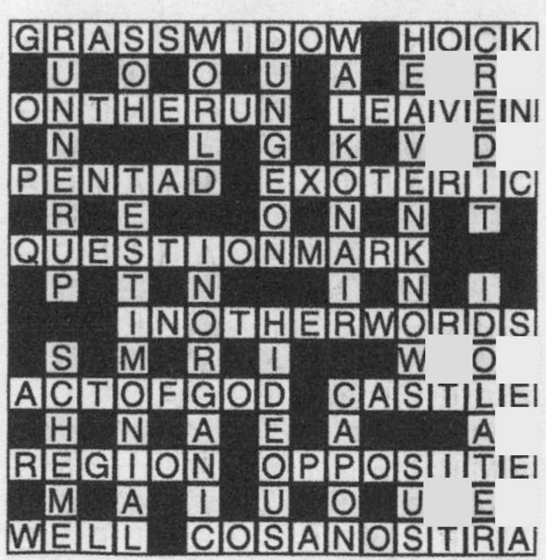

SOLUTIONS

21

G	A	F	F	E	R	T	A	P	E		L	E	N	S
U		U		A		R		E				X		O
F	I	N	I	S	T	E	R	R	E		S	P	A	M
F		K		Y		A		C		D		E		B
			C	O	L	D	S	H	O	U	L	D	E	R
S		K		N				A		T		I		E
M	I	N	U	T	E	M	A	N		C	U	T	I	N
A		O		H		E		C		H		E		E
T	A	B	L	E		T	H	E	S	A	U	R	U	S
T		K		E		H				U		S		S
E	V	E	R	Y	M	A	N	J	A	C	K			
R		R		E		D		O		T		M		S
I	N	R	O		T	O	U	R	N	I	Q	U	E	T
N		I				N		U		O		L		A
G	E	E	K		T	E	R	M	I	N	A	L	L	Y

22

H	A	N	G	O	V	E	R		G			K		W
	R		U		E		E	X	A	M	I	N	E	E
I	M	P	I		R		P		M			E		A
	A		L	I	T	T	L	E	B	O	P	E	E	P
	T		D		E		A		O		R			O
S	U	P	E	R	B		C	U	L	L	O	D	E	N
	R		N		R		E				P			R
M	E	S	S	I	A	H		D	E	B	U	S	S	Y
O			T				D		S		P		E	
N	O	V	E	M	B	E	R		T	A	T	A	M	I
M			R		R		A		R		H		I	
O	W	E	N	W	I	N	G	R	A	V	E		N	
U		R			D		N		G		B	R	A	G
T	R	I	P	L	A	N	E		O		A		R	
H		E			L		T	A	N	T	R	I	S	M

23

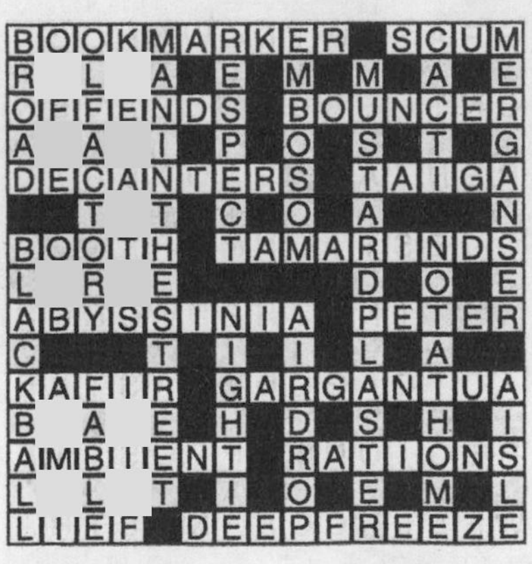

B	O	O	K	M	A	R	K	E	R		S	C	U	M
R		L		A		E		M		M		A		E
O	F	F	E	N	D	S		B	O	U	N	C	E	R
A		A		I		P		O		S		T		G
D	E	C	A	N	T	E	R	S		T	A	I	G	A
		T		T		C		O		A				N
B	O	O	T	H		T	A	M	A	R	I	N	D	S
L		R		E						D		O		E
A	B	Y	S	S	I	N	I	A		P	E	T	E	R
C				T		I		I		L		A		
K	A	F	I	R		G	A	R	G	A	N	T	U	A
B		A		E		H		D		S		H		I
A	M	B	I	E	N	T		R	A	T	I	O	N	S
L		L		T		I		O		E		M		L
L	I	E	F		D	E	E	P	F	R	E	E	Z	E

24

S	N	A	P	P	Y		H	A	I	R	L	I	N	E
A		S		I				L		E		N		N
N	O	H	O	P	E	R		L	U	C	I	F	E	R
D		E		E		O		I		I		O		O
P	A	N	E	L		U	N	N	A	T	U	R	A	L
A				I		G		G		E		M		
P	A	R	E	N	T	H	O	O	D		F	A	C	T
E		E		E		A		O		A		N		E
R	U	S	H		U	N	E	D	U	C	A	T	E	D
		E		B		D		T		H				D
P	O	T	P	O	U	R	R	I		I	M	P	L	Y
A		T		R		E		M		L		U		B
I	L	L	Y	R	I	A		E	C	L	I	P	S	E
R		E		O		D				E		I		A
S	I	D	E	W	A	Y	S		O	S	T	L	E	R

25

S	L	I	P	U	P		C	L	A	M	B	A	K	E
	O		O		I		H		B		A		N	
C	O	N	C	R	E	T	E		S	I	G	N	A	L
	K		H				E		E		L		V	
B	A	N	A	N	A	S	K	I	N		A	P	E	X
	L		R		S				T		D			
B	I	R	D		S	C	A	R	E	D	Y	C	A	T
	K				O		W		E				R	
V	E	R	M	I	C	E	L	L	I		C	A	G	E
			U		I				S		E		E	
O	C	H	E		A	Q	U	A	M	A	R	I	N	E
	R		Z		T		N				A		T	
B	A	N	Z	A	I		C	H	O	W	M	E	I	N
	W		I		V		U		A		I		N	
P	L	A	N	G	E	N	T		R	O	C	K	E	T

26

N	E	F	A	R	I	O	U	S		S	H	A	R	P
E		R		A		L		T		A		G		E
G	L	I	M	P	S	E		A	I	L	M	E	N	T
U		E		I		A		R		L		O		U
S	Y	N	O	D		S	O	R	R	O	W	F	U	L
		D		I		T		Y		W		D		A
A	X	L	E	T	R	E	E				M	I	E	N
R		Y		Y		R		V		M		S		C
T	R	I	P				M	E	D	I	O	C	R	E
I		S		M		H		N		N		R		
F	A	L	S	E	H	O	O	D		S	H	E	L	L
I		A		M		W		E		T		T		E
C	O	N	C	O	R	D		T	A	R	N	I	S	H
E		D		R		A		T		E		O		A
R	I	S	K	Y		H	E	A	D	L	I	N	E	R

27

W	H	I	T	E	S		C	L	U	B	L	A	N	D
	E		U		L		O			R		B		A
P	A	N	T	H	E	O	N		L	I	V	E	R	Y
	R		E		E		F			D		D		L
P	H	I	L	I	P	P	I		B	L	E	N	D	E
	E		A		O		D			E		E		W
	A		R	E	V	I	E	W	E	D		G		I
F	R	A	Y		E		N		V		R	O	W	S
L		V		P	R	A	C	T	I	C	E		I	
E		E		E			E		L		C		S	
T	U	R	B	A	N		T	I	D	E	O	V	E	R
C		A		K			R		O		I		A	
H	A	G	G	I	S		I	D	I	O	L	E	C	T
E		E		E			C		N		E		R	
R	O	D	E	R	I	C	K		G	A	D	G	E	T

28

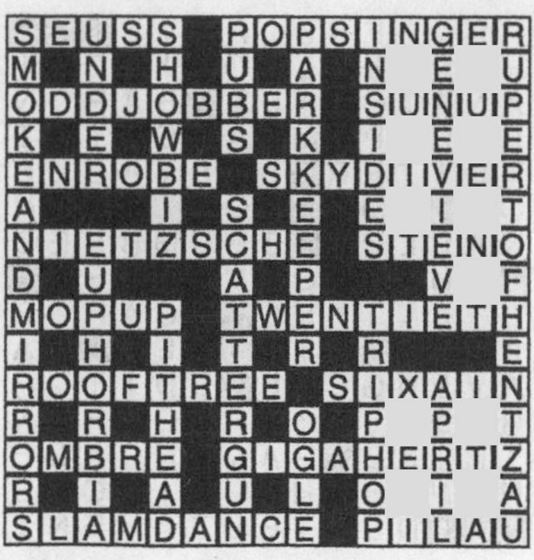

S	E	U	S	S		P	O	P	S	I	N	G	E	R
M		N		H		U		A		N		E		U
O	D	D	J	O	B	B	E	R		S	U	N	U	P
K		E		W		S		K		I		E		E
E	N	R	O	B	E		S	K	Y	D	I	V	E	R
A				I		S		E		E		I		T
N	I	E	T	Z	S	C	H	E		S	T	E	N	O
D		U				A		P				V		F
M	O	P	U	P		T	W	E	N	T	I	E	T	H
I		H		I		T		R		R				E
R	O	O	F	T	R	E	E		S	I	X	A	I	N
R		R		H		R		O		P		P		T
O	M	B	R	E		G	I	G	A	H	E	R	T	Z
R		I		A		U		L		O		I		A
S	L	A	M	D	A	N	C	E		P	I	L	A	U

SOLUTIONS

29

30

31

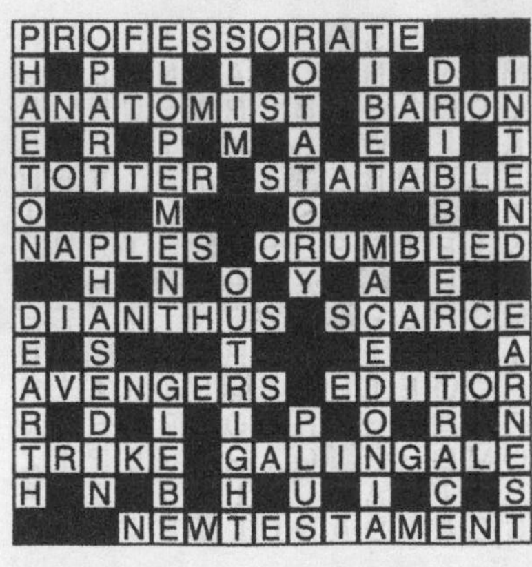

32

33

W	R	I	T	E	O	F	F		I	N	D	I	G	O
	E		A		S		E		R		O		A	
	J	O	B	S	C	O	M	F	O	R	T	E	R	
	O		L		A		O		N				O	
G	I	D	E	O	N		R	O	S	S	E	T	T	I
	C		T				A		I		L		T	
B	E	D	E	V	I	L	L	E	D		A	B	E	L
			N		M				E		S			
O	M	A	N		A	C	C	U	S	A	T	I	V	E
	O		I		G		H				I		I	
C	H	A	S	T	I	S	E		S	E	C	T	O	R
	I				N		L		N		B		L	
	C	O	M	P	A	S	S	I	O	N	A	T	E	
	A		A		R		E		R		N		N	
S	N	A	P	P	Y		A	N	T	I	D	O	T	E

34

S	P	R	I	N	G	E	R		F			E		I
	O		M		L		E	V	E	R	Y	M	A	N
T	R	I	P	L	E	T	S		A			I		A
	T		O		N		P	A	T	H	O	G	E	N
	I		R				O		H			R		D
P	E	S	T	I	L	E	N	C	E		S	A	G	O
	R		E		A		D		R			T		U
B	E	L	D	A	M	E		R	I	C	H	E	S	T
O		A			E		A		N		A		C	
N	O	N	E		N	A	V	I	G	A	T	I	O	N
A		G			T		E				C		R	
F	O	U	N	T	A	I	N		O		H		P	
I		I			B		G	A	R	D	E	N	I	A
D	I	S	S	O	L	V	E		A		R		O	
E		H			E		R	E	L	A	Y	I	N	G

35

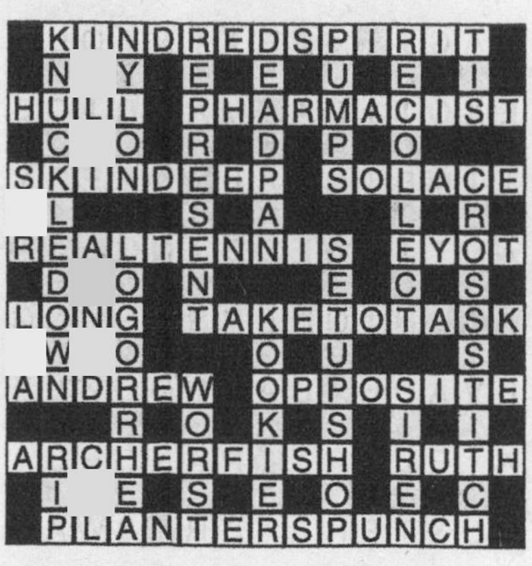

36

C	Y	P	R	E	S	S		B	O	S	W	E	L	L
A		O		L		K		A		H		S		A
R	E	P	E	L	L	E	N	T		A	S	P	E	N
D		U		E		L		H		R		A		D
		L	A	N	C	E	C	O	R	P	O	R	A	L
S		A				T		S		N		T		O
A	B	R	A	S	I	O	N		H	E	R	O	I	C
U				H		N		F		S				K
E	S	P	R	I	T		C	O	N	S	P	I	R	E
R		R		P		H		R				N		D
K	N	O	W	W	H	A	T	S	W	H	A	T		
R		F		R		N		O		Y		E		M
A	D	A	G	E		G	L	O	W	I	N	G	L	Y
U		N		C		A		T		N		E		T
T	R	E	K	K	E	R		H	O	G	A	R	T	H

SOLUTIONS

37

38

39

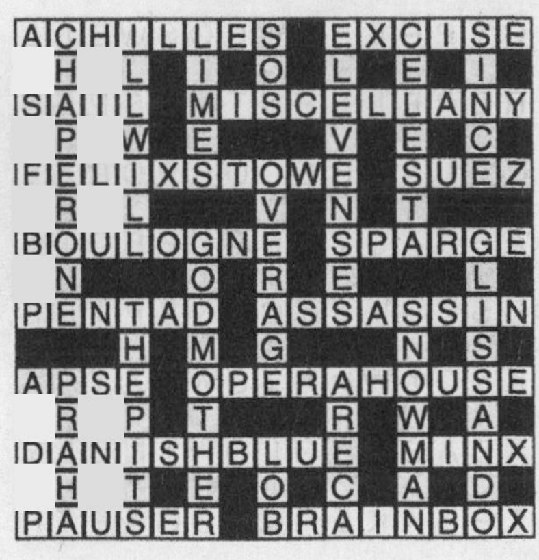

40

SOLUTIONS

41

C	L	A	S	S	I	F	Y		I	S	L	A	N	D
H		D		P		U		P		A		M		I
A	D	O	R	E		C	O	L	O	N	N	A	D	E
R		P		C		H		U		A		Z		S
I	N	T	R	I	N	S	I	C		N	I	E	C	E
S				A		I		K		D		M		L
M	A	T	I	L	D	A		Y	O	R	K	E	R	
A		W		P						E		N		D
	C	O	U	L	I	S		S	H	A	T	T	E	R
I		A		E		T		A		S				U
C	O	P	R	A		O	U	T	O	F	H	A	N	D
I		E		D		R		I		A		G		G
C	A	N	T	I	L	E	N	A		U	S	A	G	E
L		N		N		Y		T		L		T		R
E	R	Y	N	G	O		L	E	A	T	H	E	R	Y

42

S	U	P	E	R	N	O	V	A		M	O	T	I	F
L		U		A		R		T		O		H		I
A	L	L	O	V	E	R	T	H	E	P	L	A	C	E
N		L		E		A		A		P		T		L
G	R	O	U	N	D		U	N	V	E	R	S	E	D
		N		O		E		A		T		M		F
W	O	E	F	U	L	N	E	S	S		C	O	D	A
I		S		S		G		I		S		R		R
L	A	S	T		A	L	T	A	R	P	I	E	C	E
L		O		O		I		N		U		L		
P	A	C	I	F	I	S	M		A	R	T	I	S	T
O		K		F		H		B		G		K		I
W	I	S	H	S	O	M	E	O	N	E	W	E	L	L
E		U		E		A		O		A		I		E
R	E	P	O	T		N	U	M	E	R	A	T	O	R

43

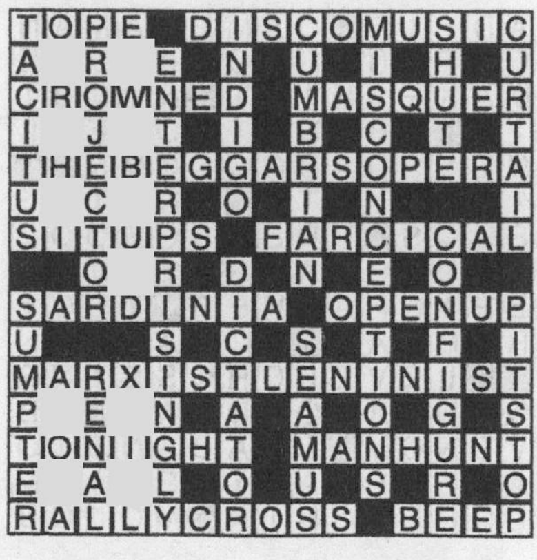

T	O	P	E		D	I	S	C	O	M	U	S	I	C
A		R		E		N		U		I		H		U
C	R	O	W	N	E	D		M	A	S	Q	U	E	R
I		J		T		I		B		C		T		T
T	H	E	B	E	G	G	A	R	S	O	P	E	R	A
U		C		R		O		I		N				I
S	I	T	U	P	S		F	A	R	C	I	C	A	L
		O		R		D		N		E		O		
S	A	R	D	I	N	I	A		O	P	E	N	U	P
U				S		C		S		T		F		I
M	A	R	X	I	S	T	L	E	N	I	N	I	S	T
P		E		N		A		A		O		G		S
T	O	N	I	G	H	T		M	A	N	H	U	N	T
E		A		L		O		U		S		R		O
R	A	L	L	Y	C	R	O	S	S		B	E	E	P

44

U	N	O	P	P	O	S	E	D		B	O	M	B	E
L		V		O		H		I		I		A		X
C	H	A	R	L	I	E		M	I	D	D	L	E	C
E		T		Y		E				E		A		E
R	E	I	G	N		P	R	A	C	T	I	C	A	L
		O		E		F		T				H		L
F	U	N		S	T	A	F	F	O	F	L	I	F	E
O				I		R		I		U				N
R	E	A	R	A	D	M	I	R	A	L		M	A	T
E		V				E		S		L		A		
S	M	E	L	L	A	R	A	T		M	O	R	A	L
I		R		O				H		A		C		I
G	R	A	N	D	M	A		A	I	R	P	O	R	T
H		G		G		S		N		K		N		H
T	H	E	M	E		S	I	D	E	S	W	I	P	E

SOLUTIONS

45

F	O	L	L	O	W	M	Y	L	E	A	D	E	R	
O		A		A		O		A		P		X		C
U	P	P	E	R	M	O	S	T		R	E	C	T	O
R		W		E		N		E		I		E		N
L	A	I	R	D		F	O	L	K	L	O	R	I	C
E		N				I		Y		F		P		R
T	I	G	H	T	E	S	T		B	O	T	T	L	E
T				E		H		A		O				T
E	M	B	A	L	M		A	P	P	L	E	P	I	E
R		I		E		T		P				E		J
W	O	R	K	P	I	E	C	E		B	A	N	T	U
O		E		H		A		N		E		S		N
R	A	T	I	O		S	A	D	D	E	N	I	N	G
D		T		T		E		I		C		O		L
	L	A	B	O	U	R	E	X	C	H	A	N	G	E

46

E	S	P	A	L	I	E	R		F	A	T	H	O	M
S		A		E		M		T		N		E		A
C	A	N	T	E	R	B	U	R	Y	T	A	L	E	S
A		E		W		E		U		W		L		S
P	O	L	L	A	R	D		E	Y	E	S	O	R	E
E		L		R				B		R				U
	W	I	N	D	F	A	L	L		P	A	R	I	S
A		S				L		U				A		E
M	E	T	A	L		A	L	E	H	O	U	S	E	
I				U		C				U		P		R
C	U	R	A	C	O	A		R	A	T	A	B	L	E
A		U		I		R		I		P		E		D
B	E	N	E	F	I	T	O	F	C	L	E	R	G	Y
L		E		E		E		L		A		R		E
E	N	S	U	R	E		D	E	W	Y	E	Y	E	D

47

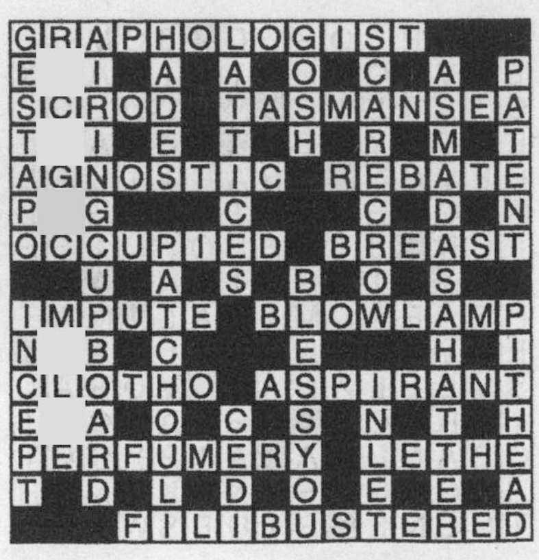

G	R	A	P	H	O	L	O	G	I	S	T			
E		I		A		A		O		C		A		P
S	C	R	O	D		T	A	S	M	A	N	S	E	A
T		I		E		T		H		R		M		T
A	G	N	O	S	T	I	C		R	E	B	A	T	E
P		G				C				C		D		N
O	C	C	U	P	I	E	D		B	R	E	A	S	T
		U		A		S		B		O		S		
I	M	P	U	T	E		B	L	O	W	L	A	M	P
N		B		C				E				H		I
C	L	O	T	H	O		A	S	P	I	R	A	N	T
E		A		O		C		S		N		T		H
P	E	R	F	U	M	E	R	Y		L	E	T	H	E
T		D		L		D		O		E		E		A
			F	I	L	I	B	U	S	T	E	R	E	D

48

A	D	L	I	B		C	A	M	E	R	A	M	A	N
D		E		U		H		O		E		E		E
M	A	K	E	S	H	I	F	T		N	I	M	B	I
I				B		N		T		O		E		L
N	E	L	S	O	N	S	C	O	L	U	M	N		
I		I		Y		T				N		T		F
S	U	Q	S		P	R	E	P	S	C	H	O	O	L
T		U		G		A		R		E		M		Y
E	P	I	G	R	A	P	H	E	R		Y	O	G	I
R		D		A				O		A		R		N
		L	E	V	E	L	C	R	O	S	S	I	N	G
O		U		A		O		D		W				B
V	E	N	O	M		B	R	A	B	A	N	T	I	O
A		C		E		A		I		R		A		M
L	O	H	E	N	G	R	I	N		M	C	J	O	B

SOLUTIONS

49

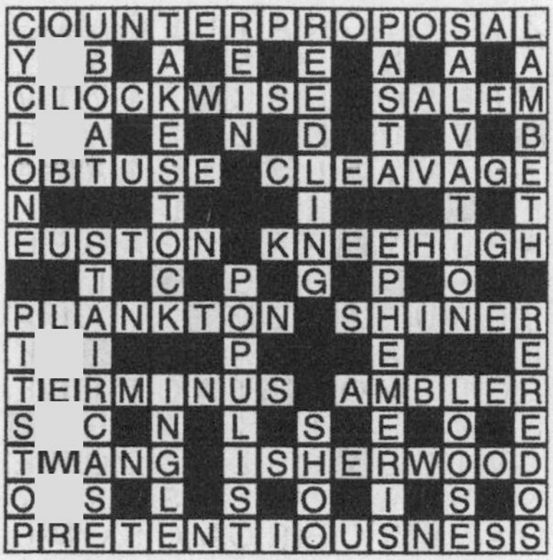

50

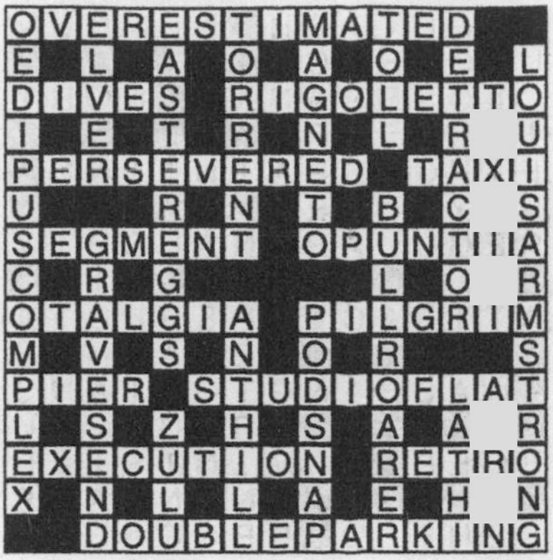

51

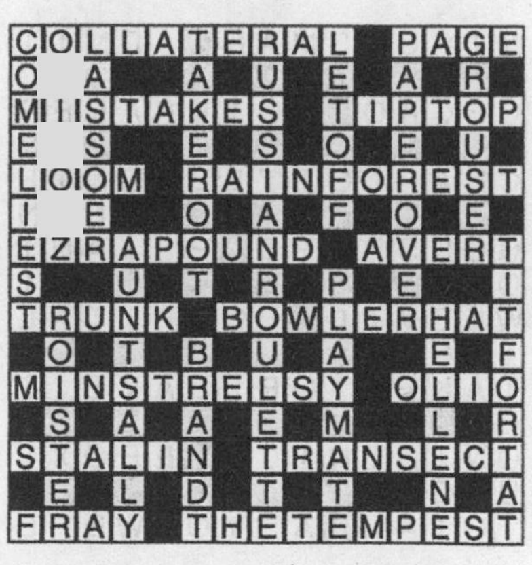

52

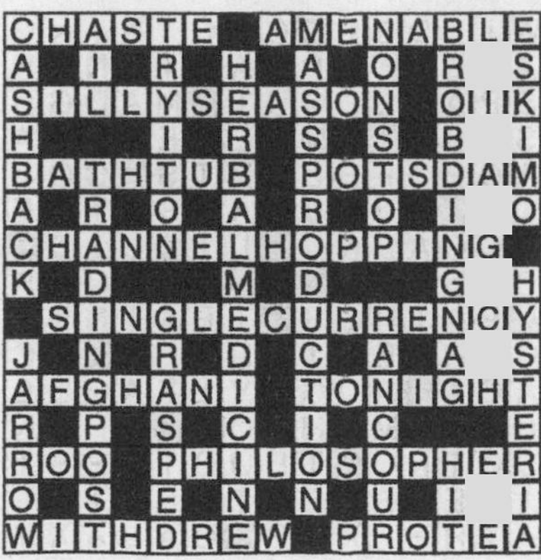

SOLUTIONS

53

```
LIGHTISH DUMDUM
U O H C  I O P 
CLOSERANKS RORY
I D B R  E N I 
FINGERPAINTING 
E I L E  G N H 
REGULAR NAUGHTY
  H  R   G  I  
ATTRACT PERUGIA
 O E H  E A H S
 WASHINGTONIANS
 R E B  R G L U
BOER ADMINISTER
 P V L  N N A E
BEHELD BEGGARED
```

54

```
ZILLIONAIRE WIG
A I N E N X A I
PIPSQUEAK PAYTV
  P U D E L W E
TRIVIAL DIORAMA
E   R E   D R N
ABBEY FATHEADED
R E   U O   L T
JERUSALEM PHYLA
E Y A   A I   K
ROLLMOP TACTILE
K L E I I A R  
IVIES TELESCOPE
N U E C L S K Y
GAM XCHROMOSOME
```

55

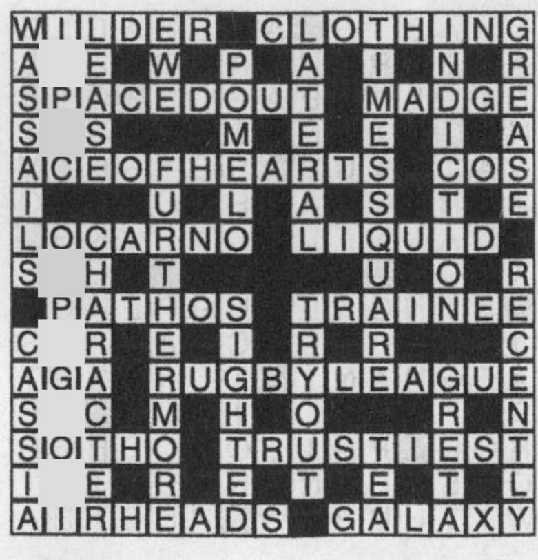

```
WILDER CLOTHING
A E W P A I N R
SPACEDOUT MADGE
S S   M E E I A
ACEOFHEARTS COS
I   U L A S T E
LOCARNO LIQUID 
S H T     U O R
 PATHOS TRAINEE
C R E I R R   C
AGA RUGBYLEAGUE
S C M H O   R N
SOTHO TRUSTIEST
I E R E T E T L
AIRHEADS GALAXY
```

56

```
MINUTEMAN ADMIT
A E R I U P O O
GECKO NOTEPAPER
I K L S M A   M
CHARLOTTEBRONTE
  N O R G E E N
GODSPEED PLIANT
O N   L C   R I
WEEVIL HOSPITAL
I C N G L L H  
TAKEITORLEAVEIT
H   T D I T B H
ONTHINICE EPOXY
U O A V R A N M
TOTAL ASYOUWERE
```

57

```
P A R E X C E L L E N C E ■ ■
L   U   E ■ X ■ E ■ U ■ X ■ A
A K R O N ■ C H A R L A T A N
S   A   O ■ I ■ F ■ L ■ R ■ O
T E L E P A T H I C ■ J E S T
E ■ ■ ■ H ■ E ■ E ■ E ■ M ■ H
R E S P O N D ■ R E Q U I R E
O   C   B ■ ■ ■ ■ ■ U ■ S ■ R
F A I R I E S ■ F R A N T I C
P   N   C ■ T ■ I ■ T ■ ■ ■ O
A R T Y ■ G O O D F O R Y O U
R   I   I ■ I ■ E ■ R ■ I ■ N
I L L O G I C A L ■ I N E P T
S ■ L   O ■ A ■ I ■ A ■ L ■ R
■ ■ A P R I L F O O L S D A Y
```

58

```
R I C H M O N D ■ G H E T T O
E ■ H ■ A ■ O ■ ■ ■ A ■ H   N
P R I O R I T Y ■ G R E E N E
A ■ L ■ M ■ I ■ M ■ S ■ O   B
S A L S A ■ F O O L H A R D Y
T ■ I ■ L ■ I ■ U ■ ■ ■ I   O
■ ■ ■ D A N C I N G Q U E E N
P ■ A ■ D ■ A ■ T ■ U ■ S ■ E
R E D L E T T E R D A Y ■ ■ ■
O ■ O ■ ■ ■ I ■ A ■ D ■ H ■ S
P E P P E R O N I ■ R E A C T
O ■ T ■ T ■ N ■ N ■ I ■ N   E
S L I G H T ■ F I L L E D U P
E ■ O ■ E ■ ■ ■ E ■ L ■ E   P
R A N K L E ■ T R U E B L U E
```

59

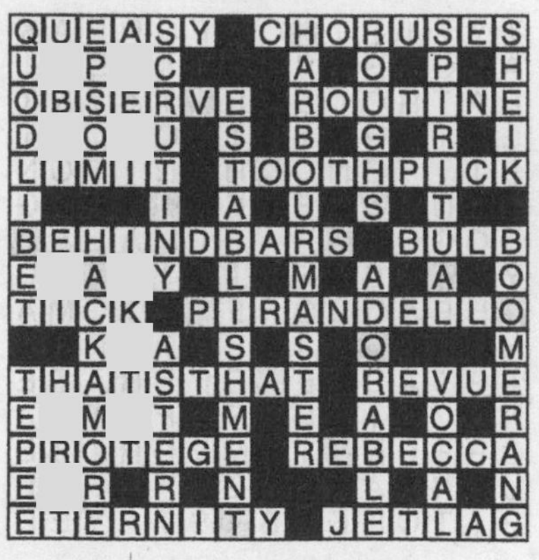

```
Q U E A S Y ■ C H O R U S E S
U   P   C ■ ■ ■ A ■ O ■ P ■ H
O B S E R V E ■ R O U T I N E
D   O   U ■ S ■ B ■ G ■ R ■ I
L I M I T ■ T O O T H P I C K
I ■ ■ ■ I ■ A ■ U ■ S ■ T ■ ■
B E H I N D B A R S ■ B U L B
E   A   Y ■ L ■ M ■ A ■ A ■ O
T I C K ■ P I R A N D E L L O
■ ■ K   A ■ S ■ S ■ O ■ ■ ■ M
T H A T I S T H A T ■ R E V U E
E   M   T ■ M ■ E ■ A ■ O ■ R
P R O T E G E ■ R E B E C C A
E   R ■ R ■ N ■ ■ ■ L ■ A ■ N
E T E R N I T Y ■ J E T L A G
```

60

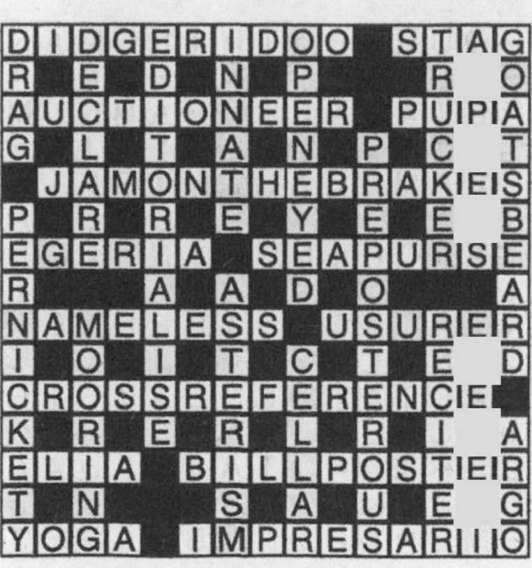

```
D I D G E R I D O O ■ S T A G
R ■ E ■ D ■ N ■ P ■ ■ ■ R   O
A U C T I O N E E R ■ P U P A
G ■ L ■ T ■ A ■ N ■ P ■ C   T
■ J A M O N T H E B R A K E S
P ■ R ■ R ■ E ■ Y ■ E ■ E   B
E G E R I A ■ S E A P U R S E
R ■ ■ ■ A ■ A ■ D ■ O ■ ■ ■ A
N A M E L E S S ■ U S U R E R
I ■ O ■ I ■ T ■ C ■ T ■ E   D
C R O S S R E F E R E N C E ■
K ■ R ■ E ■ R ■ L ■ R ■ I   A
E L I A ■ B I L L P O S T E R
T ■ N ■ ■ ■ S ■ A ■ U ■ E   G
Y O G A ■ I M P R E S A R I O
```

SOLUTIONS

61

C	O	G	N	A	C		S	C	R	A	M	B	L	E
	P		E		R		U		E		A		E	
C	A	U	T	I	O	U	S		C	O	R	R	A	L
	Q		H		M		P		A				V	
E	U	R	E	K	A		E	X	P	O	R	T	E	R
	E		R		G		N				I		O	
			L	O	N	G	D	R	A	W	N	O	U	T
	C		A		O				E		G		T	
R	U	N	N	I	N	G	C	O	S	T	S			
	R		D				O		C		P		V	
P	A	R	S	I	F	A	L		H	E	A	V	E	N
	T				O		L		Y		N		N	
H	I	L	A	R	Y		A	T	L	A	N	T	I	S
	V		I		E		G		U		E		C	
G	E	N	D	A	R	M	E		S	U	R	R	E	Y

62

S	U	I	T		S	P	O	O	N	E	R	I	S	M
T		D		O		A		R		Y		N		E
A	R	A	U	C	A	R	I	A		E	M	C	E	E
L		H		T		S		N		B		O		T
A	V	O	C	A	D	O		G	A	R	A	G	E	
C				V		N				O		N		V
T	H	E	W	O	M	A	N	I	N	W	H	I	T	E
I		A				G		M				T		N
T	H	R	E	E	M	E	N	I	N	A	B	O	A	T
E		N		N				T		B				I
	J	E	T	L	A	G		A	B	Y	S	M	A	L
B		S		A		R		T		D		A		A
E	S	T	E	R		E	P	I	T	O	M	I	S	T
A		L		G		E		O		S		Z		O
U	N	Y	I	E	L	D	I	N	G		W	E	A	R

63

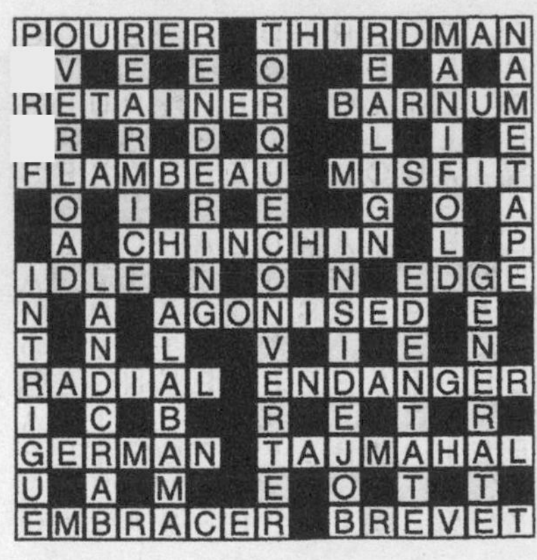

64

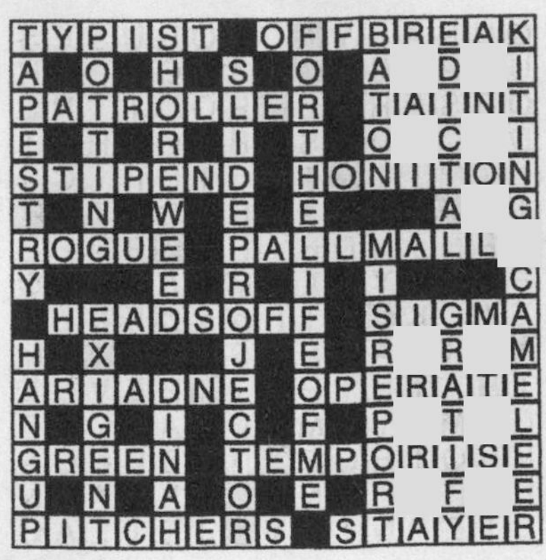

65

C	U	T	G	L	A	S	S		S	U	B	W	A	Y
H		A		E		C			H		A		I	
A	I	R		G	L	O	U	C	E	S	T	E	R	
M		S		A		U			L		T		S	
P	R	I	S	T	I	N	E		T	I	E	P	I	N
A		E		O		D			E		R		D	
G	Y	R	E		P	R	O	C	R	U	S	T	E	S
N			X			E		H			E			H
E	P	I	C	Y	C	L	O	I	D		A	C	E	R
	A		E		U			P		C		H		I
A	N	K	L	E	T		L	O	C	A	T	I	O	N
	D		L		L			L		U		C		K
	O	P	E	R	A	G	L	A	S	S		A	D	A
	R		N		S			T		E		G		G
P	A	S	T	I	S		C	A	P	R	I	O	L	E

66

R	O	B	U	S	T		A	B	S	T	R	A	C	T
A		O		U		T		A		I		N		A
R	U	N	A	C	R	O	S	S		N	E	G	U	S
I		E		C		M		I		T		E		K
N	A	R	R	O	W	M	I	N	D	E	D	L	Y	
G				U		Y				R		C		A
T	A	P	E	R		R	E	P	E	N	T	A	N	T
O		A				O		H				K		T
G	A	L	L	A	N	T	R	Y		C	R	E	P	E
O		M		N				S		H				N
	A	I	R	C	O	N	D	I	T	I	O	N	E	D
B		S		I		Y		C		N		O		A
L	I	T	H	E		A	G	A	M	E	M	N	O	N
U		R		N		L		L		S		E		C
E	G	Y	P	T	I	A	N		S	E	T	T	L	E

67

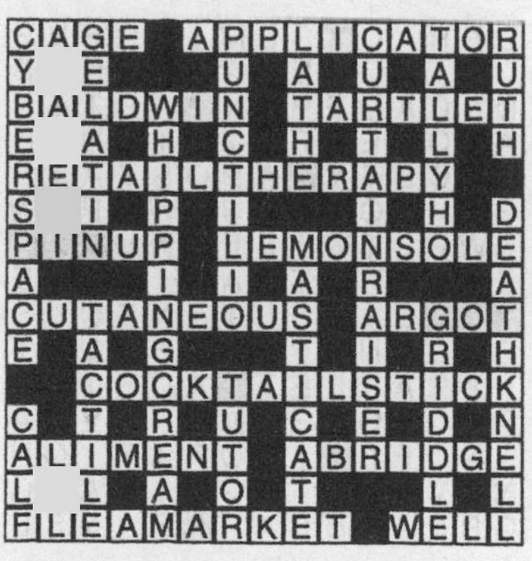

C	A	G	E		A	P	P	L	I	C	A	T	O	R
Y		E				U		A		U		A		U
B	A	L	D	W	I	N		T	A	R	T	L	E	T
E		A		H		C		H		T		L		H
R	E	T	A	I	L	T	H	E	R	A	P	Y		
S		I		P		I				I		H		D
P	I	N	U	P		L	E	M	O	N	S	O	L	E
A				I		I		A		R				A
C	U	T	A	N	E	O	U	S		A	R	G	O	T
E		A		G				T		I		R		H
		C	O	C	K	T	A	I	L	S	T	I	C	K
C		T		R		U		C		E		D		N
A	L	I	M	E	N	T		A	B	R	I	D	G	E
L		L		A		O		T				L		L
F	L	E	A	M	A	R	K	E	T		W	E	L	L

68

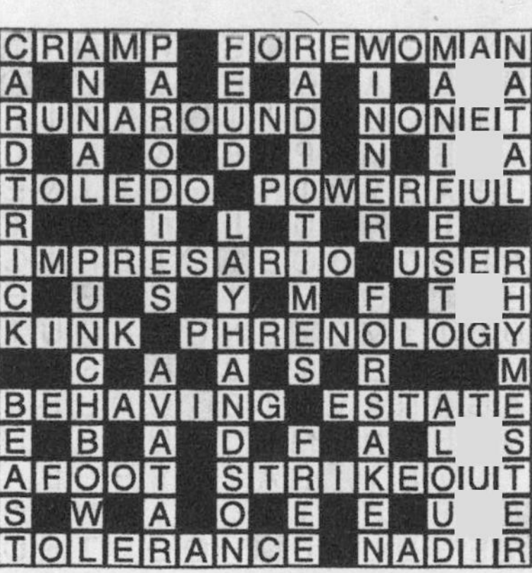

C	R	A	M	P		F	O	R	E	W	O	M	A	N
A		N		A		E		A		I		A		A
R	U	N	A	R	O	U	N	D		N	O	N	E	T
D		A		O		D		I		N		I		A
T	O	L	E	D	O		P	O	W	E	R	F	U	L
R				I		L		T		R		E		
I	M	P	R	E	S	A	R	I	O		U	S	E	R
C		U		S		Y		M		F		T		H
K	I	N	K		P	H	R	E	N	O	L	O	G	Y
		C		A		A		S		R				M
B	E	H	A	V	I	N	G		E	S	T	A	T	E
E		B		A		D		F		A		L		S
A	F	O	O	T		S	T	R	I	K	E	O	U	T
S		W		A		O		E		E		U		E
T	O	L	E	R	A	N	C	E		N	A	D	I	R

SOLUTIONS

69

P	A	T	E	L	L	A	E		C	A	R	P	E	L
A		A		U		L		E		B		L		A
P	I	K	E	S	T	A	F	F		D	I	A	R	Y
A		E		T		R		F		U		T		R
	TI	S	A	R		M	A	L	A	C	H	I	T	E
B		O		I				U		T		N		A
A	ID	M	I	N	I	S	T	E	R		C	U	R	D
T		E		G		P		N		G		M		E
T	U	B	E		P	O	T	T	E	R	S	B	A	R
L		E		P		R				E		L		S
E	R	A	D	I	C	A	T	E		A	M	O	S	
D		T		G		D		L		T		N		B
O	P	I	N	E		I	C	E	L	A	N	D	E	R
R		N		O		C		N		P		E		E
E	N	G	I	N	E		C	A	R	E	S	S	E	D

70

C	L	E	M	A	T	I	S		S	H	U	T	U	P
	E		I		I		Y		I		N		N	
	G	E	N	E	R	A	L	S	T	R	I	K	E	
	H		D		E		L		A				A	
F	O	R	B	A	D		A	N	T	I	D	O	T	E
	R		L				B		U		I		E	
I	N	N	O	V	A	T	I	O	N		S	U	N	K
			W		U				G		T			
T	A	X	I		D	I	S	H	A	B	I	L	L	E
	L		N		I		T				N		E	
P	E	I	G	N	O	I	R		D	E	C	O	C	T
	M				T		O		I		T		T	
	B	O	O	K	A	T	B	E	D	T	I	M	E	
	I		A		P		E		S		O		R	
E	C	U	R	I	E		S	E	T	E	N	A	N	T

71

A	L	C	O	P	O	P		S	I	P	H	O	N	
N		E		A		O		H		E		U		
T	I	M	E	L	I	M	I	T		C	U	T	I	S
I		B		M		P		E		K		L		L
C	R	A	Z	Y		I	N	T	E	S	T	I	N	E
O		L				D		L		N		E		I
A	G	O	O	D	J	O	B		O	I	L	R	I	G
G				O		U		I		F				H
U	L	T	I	M	O		S	N	U	F	F	O	U	T
L		I		D		I		J				P		O
A	G	G	R	A	V	A	T	E		G	R	U	F	F
N		H		N		M		C		U		L		H
T	U	T	S	I		B	A	T	H	S	H	E	B	A
		E		E		I		O		T		N		N
	E	N	B	L	O	C		R	O	O	S	T	E	D

72

73

B A L T I C S E A J A M E S
E I N H M U U T
V A Q U E R O B O N A N Z A
E U L R U T C G
L E I L A T A L K A T H O N
D S T A A
F E A S T E I N D H O V E N
E I I R C Y O C
M O R E C A M B E S I X T Y
I M C T P
N I N E T I E T H E B O O K
I A O M A R P A
S C O R P I O S K I J U M P
T M A R E C L O
S W I Z Z Y A R D S T I C K

74

R I C O C H E T G R O C E R
N V E I R W A
C A R E E R C L U E L E S S
P R E K B T
S T U B F I L I B U S T E R
A O E Y U N
W A L L A R O O B O D Y
T A D N S S E
E T O N E X A C T O R S
E C A S I A
U N M E R C I F U L N I C E
T A A A D R
B I L L Y C A N B E A C O N
O E I C L R W
A N G O R A Y I E L D I N G

75

C A S U A L S F O R E A R M
O K F M O E D O
N A Y F L A G G E D D O W N
V E I T Y S E
E N T I R E T Y S T A G E Y
R E M E U A O E
T A R A B R I N G R O U N D
R S I O T R
K N I C K K N A C K A M O S
E E I G C D A I
T H R I P S J U L I E N N E
C L T P V D V
H I G H A S A K I T E I R E
E Y N C E R S R
S U P R E M O D E S C E N T

76

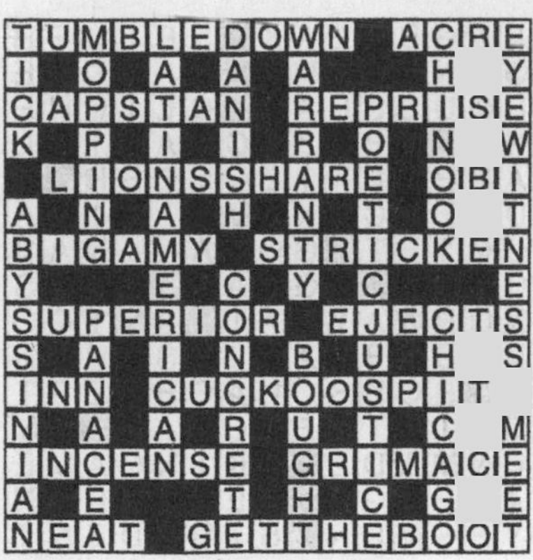

SOLUTIONS

77

L	O	C	U	S	T		I	T	C	H	I	E	S	T
I		H		P		N		H		O		M		H
B	E	A	T	I	T	U	D	E		L	O	I	R	E
I		P		L		M		G		L		L		B
D	O	E		L	I	B	E	R	T	Y	B	E	L	L
O		R				E		E		W				U
	C	O	P	P	E	R		A	P	O	S	T	L	E
S		N		A		C		T		O		H		S
C	L	E	A	R	E	R		G	O	D	O	W	N	
O				C		U		A				A		S
W	O	R	T	H	I	N	G	T	O	N		R	A	T
L		O		M		C		S		O		T		I
I	N	G	L	E		H	I	B	E	R	N	I	A	N
N		E		N		E		Y		T		N		K
G	Y	R	A	T	O	R	Y		S	H	A	G	G	Y

78

S	E	R	E	N	A	D	E	R		S	C	R	E	W
A		E		E		E		O		K		E		I
C	H	A	T	E	A	U		C	H	I	F	F	O	N
K		L		D		T				E		R		D
S	T	I	L	L		E	X	T	E	R	N	A	L	S
		S		E		R		H				I		W
B	U	T		F	L	O	R	E	S	C	E	N	C	E
E				U		N		G		A				P
S	N	O	W	L	E	O	P	A	R	D		P	I	T
E		P				M		M		E		L		
T	O	U	C	H	T	Y	P	E		T	H	E	T	A
T		L		A				I		S		U		M
I	T	E	M	I	S	E		S	C	H	E	R	Z	O
N		N		R		M		U		I		A		U
G	U	T	S	Y		U	N	P	O	P	U	L	A	R

79

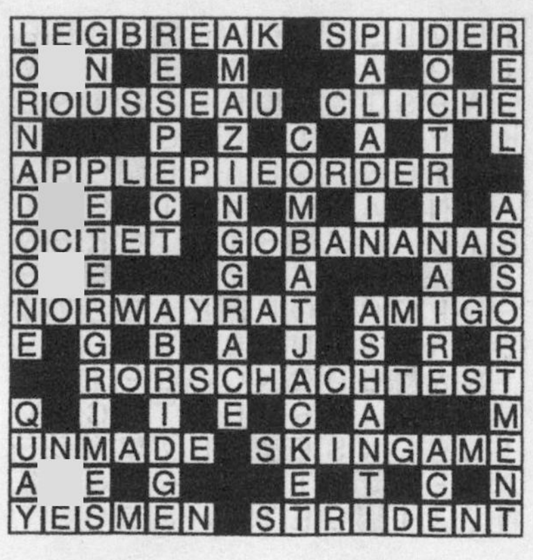

L	E	G	B	R	E	A	K		S	P	I	D	E	R
O		N		E		M				A		O		E
R	O	U	S	S	E	A	U		C	L	I	C	H	E
N				P		Z		C		A		T		L
A	P	P	L	E	P	I	E	O	R	D	E	R		
D		E		C		N		M		I		I		A
O	C	T	E	T		G	O	B	A	N	A	N	A	S
O		E				G		A				A		S
N	O	R	W	A	Y	R	A	T		A	M	I	G	O
E		G		B		A		J		S		R		R
		R	O	R	S	C	H	A	C	H	T	E	S	T
Q		I		I		E		C		A				M
U	N	M	A	D	E		S	K	I	N	G	A	M	E
A		E		G				E		T		C		N
Y	E	S	M	E	N		S	T	R	I	D	E	N	T

80

P	U	P	I	L		C	O	M	E	D	I	A	N	
O		Y		O		E		A		I		F		
S	I	L	E	N	T	M	A	J	O	R	I	T	Y	
I		O		G		E		O		A		E		L
T	E	N	S	I	O	N	A	L		C	O	R	G	I
I				T		T		I				N		F
V	I	G	O	U	R		S	C	H	M	O	O	Z	E
E		R		D		S		A		A		O		S
S	K	E	L	E	T	O	N		Q	U	I	N	C	E
I		G				M		A		R				N
G	H	O	S	T		B	A	N	D	I	C	O	O	T
N		R		W		R		N		T		X		E
	L	I	M	I	T	E	D	E	D	I	T	I	O	N
		A		C		R		X		U		D		C
	E	N	V	E	L	O	P	E		S	U	E	D	E

Test your word power to its full with these brain-teasing cryptic crosswords from The Times:

***The Times* Crossword Book 1**
ISBN 13: 978-0-00-710833-6

***The Times* Crossword Book 2**
ISBN 13: 978-0-00-711581-5

***The Times* Crossword Book 3**
ISBN 13: 978-0-00-712195-3

***The Times* Crossword Book 4**
ISBN 13: 978-0-00-712674-3

***The Times* Crossword Book 5**
ISBN 13: 978-0-00-714496-9

***The Times* Crossword Book 6**
ISBN 13: 978-0-00-714626-0

***The Times* Crossword Book 7**
ISBN 13: 978-0-00-716538-4

***The Times* Crossword Book 8**
ISBN 13: 978-0-00-716542-1

***The Times* Crossword Book 9**
ISBN 13: 978-0-00-719838-2

***The Times* Crossword Book 10**
ISBN 13: 978-0-00-719836-8

***The Times* Crossword Book 11**
ISBN 13: 978-0-00-721440-2

***The Times* Crossword Collection**
ISBN 13: 978-0-00-721300-9

Conceived to vex your wits and baffle your brain, try the Jumbo Cryptic Crossword challenge from The Times:

***The Times* Jumbo Cryptic Crossword Book 6**
ISBN 13: 978-0-00-710833-6

***The Times* Jumbo Cryptic Crossword Book 7**
ISBN 13: 978-0-00-711581-5